Reptiles and Amphibians of Canada

Chris Fisher, Amanda Joynt, Ronald J. Brooks

Lone Pine Publishing

First printed in 2007 10 9 8 7 6 5 4 3 2 1
Printed in China

The Publisher: Lone Pine Publishing
10145 – 81 Avenue
Edmonton, AB T6E 1W9
Website: www.lonepinepublishing.com

Library and Archives Canada Cataloguing in Publication

Fisher, Chris C. (Christopher Charles), 1970-
Reptiles and amphibians of Canada / Chris Fisher, Amanda Joynt, Ronald J. Brooks.

Includes index.
ISBN-13: 978-1-55105-279-3
ISBN-10: 1-55105-279-2

1. Reptiles--Canada. 2. Amphibians--Canada. I. Joynt, Amanda, 1977- II. Brooks, Ronald J., 1941 III. Title.

QL654.F57 2007 597.9'0971 C2007-900505-5

Editorial Director: Nancy Foulds
Project Editor: Sheila Quinlan
Editorial Support: Genevieve Boyer, Sandra Bit
Illustrations Coordinator: Carol Woo
Production Manager: Gene Longson
Book Design & Layout: Heather Markham
Production Support: Trina Koscielnuk, Willa Kung
Cover Design: Gerry Dotto
Maps: Elliot Engley
Photo Coordinator: Randy Kennedy
Digital Image Scans: Elite Lithographers Co., ColorSpace Photo-Graphics Inc.

All illustrations by Gary Ross except: Michel Poirier 26, 40, 80, 84, 88, 92, 103b, 106, 126, 136, 152, 154, 156; Ian Sheldon 44, 50, 124, 132 (page numbers refer to primary illustrations).

All photographs by photos.com

We acknowledge the financial support of the Government of Canada through the Book Publishing Industry Development Program (BPIDP) for our publishing activities.

PC: P14

Table of Contents

Acknowledgements

This project began several years ago. In the time that has elapsed since then, several people were involved in creating lists of species, writing, illustrating, drawing maps, putting together photos, editing and design and layout. We'd like to thank Chris Fisher and Amanda Joynt for their work on the lists and text. Ron Brooks at the University of Guelph was the technical advisor checking over species accounts, drawing up maps and looking over the illustrations, which were prepared by Gary Ross. Randy Kennedy supplied the photos through photos.com. We'd like to thank Sheila Quinlan for her fine edit of the text and for organizing the project through to production. Finally, in production we have Heather Markham to thank for the design and layout of the interior of the book and Gerry Dotto for cover design.

Species at a Glance • Reptiles

Snapping Turtle
22–47 cm • p. 26

Stinkpot
7–13 cm • p. 28

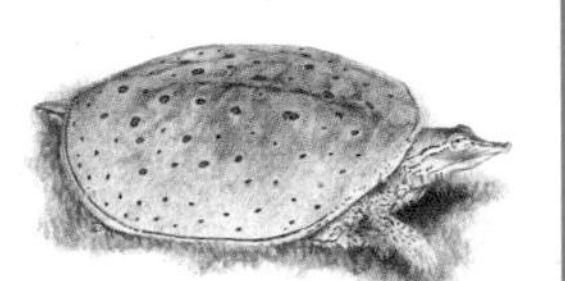

Spiny Softshell
12–42 cm • p. 30

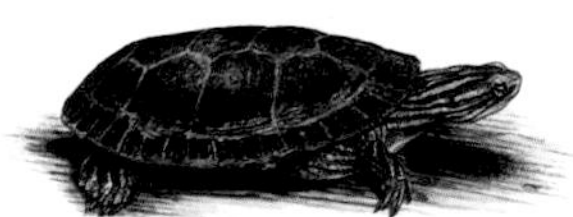

Painted Turtle
10–25 cm • p. 32

Spotted Turtle
9–14 cm • p. 34

Blanding's Turtle
18–26 cm • p. 36

Wood Turtle
16–25 cm • p. 38

Northern Map Turtle
10–27 cm • p. 40

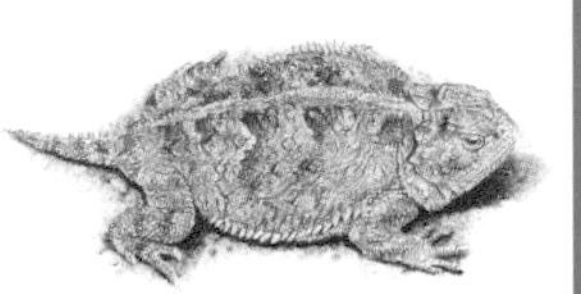

Greater Short-horned Lizard
5–11 cm • p. 42

Northern Alligator Lizard
15–25 cm • p. 44

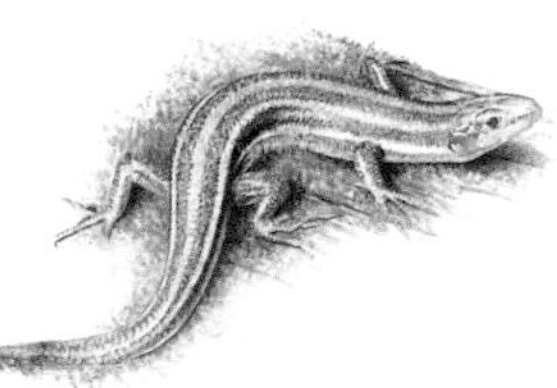

Five-lined Skink
13–21 cm • p. 46

Prairie Skink
13–22 cm • p. 48

Reptiles • Species at a Glance

Western Skink
15–18 cm • p. 50

Rubber Boa
35–80 cm • p. 52

Racer
60 cm–2 m • p. 54

Sharp-tailed Snake
20–45 cm • p. 56

Ring-necked Snake
25–60 cm • p. 58

Eastern Foxsnake
90 cm–1.7 m • p. 60

Eastern Ratsnake
1–2.6 m • p. 62

Western Hog-nosed Snake
40–90 cm • p. 64

Eastern Hog-nosed Snake
50 cm–1.2 m • p. 66

Nightsnake
30–60 cm • p. 68

Milksnake
60 cm–1.4 m • p. 70

Northern Watersnake
60 cm–1.4 m • p. 72

Smooth Greensnake
30–60 cm • p. 74

Gophersnake
90 cm–2 m • p. 76

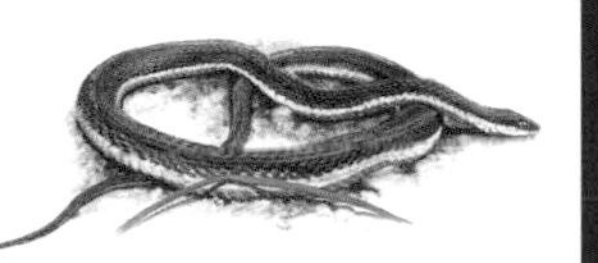
Queen Snake
40–90 cm • p. 78

DeKay's Brownsnake
30–50 cm • p. 80

Red-bellied Snake
40 cm • p. 82

Butler's Gartersnake
30–65 cm • p. 84

Terrestrial Gartersnake
45 cm–1 m • p. 86

Northwestern Gartersnake
45–95 cm • p. 88

Plains Gartersnake
50 cm–1.8 m • p. 90

Eastern Ribbonsnake
45 cm–1 m • p. 92

Common Gartersnake
85 cm–1 m • p. 94

Western Rattlesnake
60 cm–1 m • p. 96

Reptiles • Species at a Glance

SNAKES

Prairie Rattlesnake
80 cm–1.4 m • p. 98

Massasauga
60 cm–1 m • p. 100

OTHER REPTILES

Loggerhead Seaturtle
85 cm–1.2 m • p. 102

Green Seaturtle
75 cm–1.5 m • p. 103

Kemp's Ridley Seaturtle
35–75 cm • p. 103

VAGRANT SPECIES

Leatherback Seaturtle
1.3–2.4 m • p. 104

Red-eared Slider
13–29 cm • p. 105

European Wall Lizard
15–23 cm • p. 106

VAGRANT SPECIES

INTRODUCED SPECIES

Pacific Pond Turtle
12–14 cm • p. 107

Eastern Box Turtle
12–19 cm • p. 108

Pigmy Short-horned Lizard
6–8 cm • p. 109

Timber Rattlesnake
90cm–1.5 m • p. 110

EXTIRPATED SPECIES

Eastern Newt
7–14 cm • p. 112

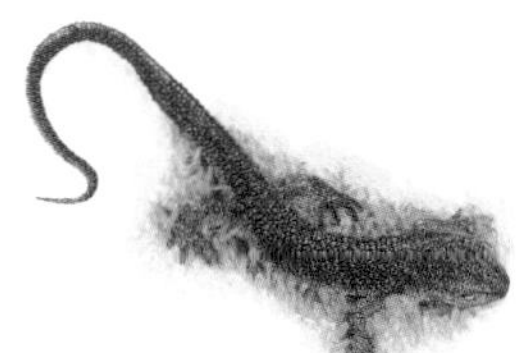
Rough-skinned Newt
12–20 cm • p. 114

Mudpuppy
20–45 cm • p. 116

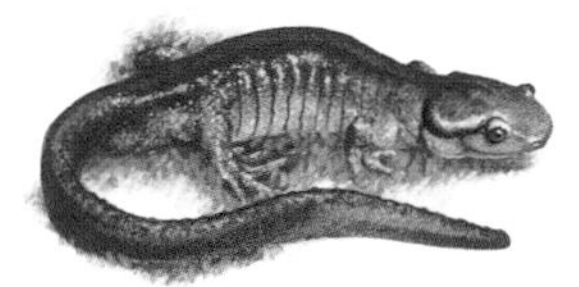
Northwestern Salamander
14–22 cm • p. 118

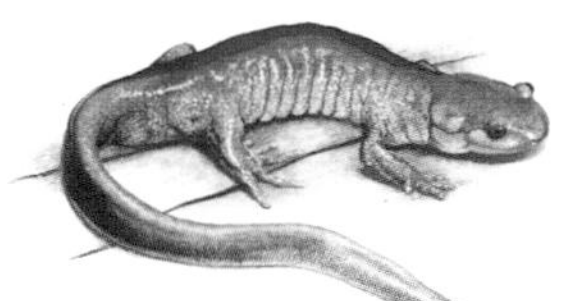
Jefferson Salamander
10–21 cm • p. 120

Blue-spotted Salamander
7–13 cm • p. 122

Long-toed Salamander
10–17 cm • p. 124

Spotted Salamander
15–25 cm • p. 126

Small-mouthed Salamander
10–20 cm • p. 128

Tiger Salamander
15–25 cm • p. 130

Coastal Giant Salamander
17–30 cm • p. 132

Wandering Salamander
7–13 cm • p. 134

SALAMANDERS

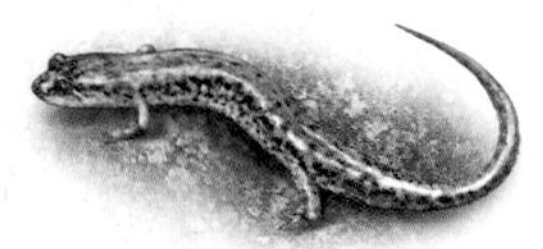

Northern Dusky Salamander
5–14 cm • p. 136

Allegheny Mountain
Dusky Salamander
7–11 cm • p. 138

Ensatina
7–15 cm • p. 140

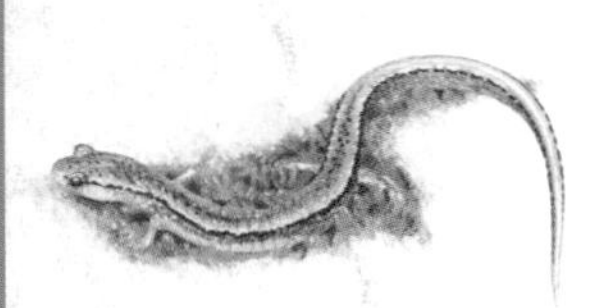

Northern Two-lined Salamander
6–12 cm • p. 142

Spring Salamander
11–21 cm • p. 144

Four-toed Salamander
5–10 cm • p. 146

Coeur d'Alene Salamander
6–12 cm • p. 148

Eastern Red-backed Salamander
6–13 cm • p. 150

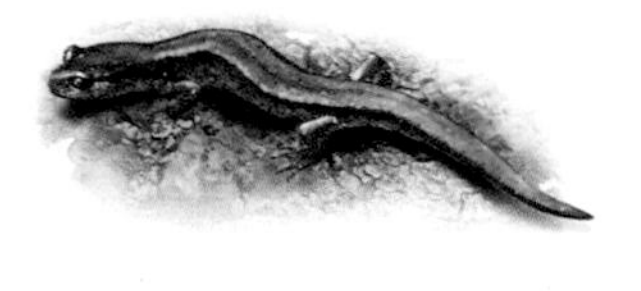

Western Red-backed Salamander
7–12 cm • p. 152

FROGS & TOADS

Rocky Mountain Tailed Frog
3–5 cm • p. 154

Coastal Tailed Frog
3–5 cm • p. 156

Plains Spadefoot
4–6 cm • p. 158

Great Basin Spadefoot
4–6 cm • p. 160

American Toad
6–11 cm • p. 162

Western Toad
6–12 cm • p. 164

Great Plains Toad
11 cm • p. 166

Canadian Toad
5–7 cm • p. 168

Fowler's Toad
4–6 cm • p. 170

Red-legged Frog
5–12 cm • p. 172

American Bullfrog
20 cm • p. 174

Green Frog
7–10 cm • p. 176

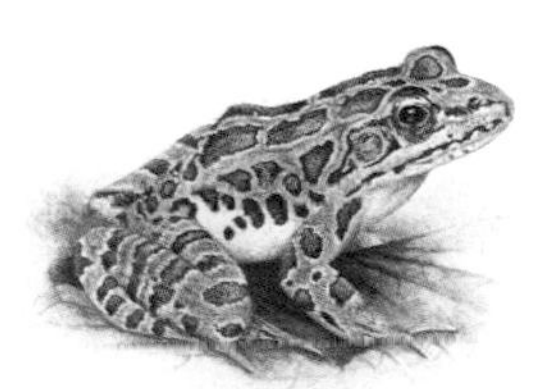

Pickerel Frog
4–9 cm • p. 178

Northern Leopard Frog
4–10 cm • p. 180

Oregon Spotted Frog
7–8 cm • p. 182

Columbia Spotted Frog
5–9 cm • p. 184

Mink Frog
5–8 cm • p. 186

Wood Frog
5–8 cm • p. 188

Northern Cricket Frog
4 cm • p. 190

Cope's Gray Treefrog
6 cm • p. 192

Gray Treefrog
6 cm • p. 194

Pacific Treefrog
3–5 cm • p. 196

Spring Peeper
3 cm • p. 198

Boreal Chorus Frog
3–4 cm • p. 200

Western Chorus Frog
3–4 cm • p. 202

Introduction

Let's face it—Canada is not exactly reptile and amphibian heaven. Elsewhere in the world, lizards commonly scramble in backyard gardens and geckoes dance on walls; here, we have to make special efforts to see lizards, and we have no geckoes. Worldwide, there are over 5500 species of amphibians and 8000 species of reptiles, but in Canada, the second largest country in the world, we have only 84 species of both—less than one percent of the world's total. However, our residents are resilient, adaptable and as cool as they come. *Reptiles and Amphibians of Canada* is a book that celebrates these wondrous animals that share our land.

Traditionally, reptiles and amphibians have been studied collectively through the science of herpetology. This grouping doesn't make any more scientific sense than studying birds and mammals together, but given their superficial similarity to one another and the small number of reptiles and amphibians in Canada in comparison to other groups of animals, the grouping seems more or less justified.

From the Beginning

Present-day reptiles and amphibians have deeply rooted family trees. Approximately 350 million years ago, fleshy-finned fish began travelling out of the water to see what the land offered. Amphibians resulted from these land pioneers and have been a huge success ever since, with over 5000 species alive today. They gradually developed

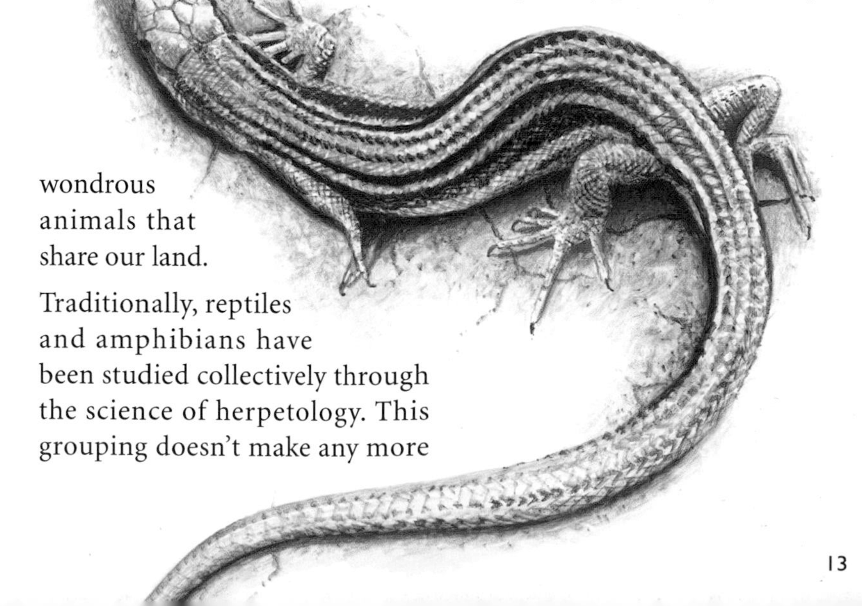

into many forms. One form, reptiles, developed about 300 million years ago. The evolution of amphibians and reptiles took place under the incubating conditions of a tropical world, one that resembled the Everglades in today's Florida. Modern-day reptiles and amphibians still flourish in tropical conditions, which explains the reduced number of species in Canada and the complete lack of them in much of the Far North.

Reptiles

Reptiles may lack the transforming flair of amphibians, but as a group they are more variable—about 8000 species—and distinct. In Canada we have representatives of three of the five living groups (orders) of reptiles: snakes, turtles and lizards. Tuataras (lizard-like animals found in New Zealand) and crocodiles are the other existing reptile orders. Several orders, such as dinosaurs and flying reptiles (pterosaurs) are extinct.

Although snakes, lizards and turtles appear dissimilar, there are unifying characteristics that reinforce their family ties. Scales are the obvious universal trait of reptiles; internal fertilization and, usually, a shelled egg are others. Of course, not all these traits are exclusive to reptiles—birds and even some fishes and mammals share these features—but they are sufficient for most people to realize when they are looking at a reptile or something else.

Unlike shell-less amphibian eggs, the eggs of reptiles are less apt to absorb external matter or dry out—meaning that reptiles must mate in a similar manner as birds and mammals so that fertilization is done internally, prior to the formation of the eggshell. Some reptiles lay eggs and build crude nests. Others, particularly species that inhabit northern areas, do not lay their eggs, but simply retain them within the body somewhat as mammals do so that the embryos develop inside the mother, who then gives birth to free-living young. Newborn reptiles enter the outside world as downsized versions of the adult. There is no larval or intermediate stage as in amphibians, and they usually grow ever larger until they die.

Amphibians

Various forms of frogs, toads and salamanders are the amphibian representatives that live in Canada. In all, there are only three groups (orders) of amphibians around today, two of which are represented in this book. Frogs and toads are the most numerous order worldwide and also here in Canada. Salamanders are much less numerous worldwide, but we have them in disproportionate numbers, a consequence that reflects their global epicentre deep in the nearby Appalachian Mountains of the eastern United States. We lack the third order completely: ceacilians are a tropical group of amphibians that superficially resemble the love child of a worm and a snake.

Amphibians do not have the ability to generate their own internal body heat as birds and mammals do, but they are not necessarily cold blooded. The preferred term is poikilothermic or ectothermic. Amphibians are not always cold; rather they tend to be the same temperature as their surroundings, and in some cases warmer because, by basking in the sun or by seeking out warm rocks, logs or water, they can raise their body temperature considerably.

The skin of amphibians is unique; it is smooth, not covered with scales, feathers or hair. Although certain skin glands can secrete an oozy protective and defensive slime, most of the time amphibian skin is soft and supple. Amphibian skin is also porous—water and

dissolved oxygen freely pass through into the blood-rich skin. Although most amphibians have lungs and breathe air, skin respiration can account for a significant portion of their oxygen uptake. Some amphibians lack lungs completely and do all their breathing through their skin.

Undeniably, the most interesting aspect of amphibians is their life cycle. The typical scenario is more similar to insects than to mammals, with an egg and an intermediate larval form. Like most animals, amphibians start off as a fertilized egg. The egg is usually deposited in water or another humid environment. The egg does not have a shell, so in the absence of water it quickly dries and dies. In most species, the tadpole emerges from the egg after a short incubation period. This larval form often does not initially resemble the adult stage. The aquatic frog and toad tadpole sprouts gills, lacks limbs and has radically different mouth structures and, subsequently, food needs from the adult. The slow and deliberate metamorphosis of a tadpole into an adult amphibian is the subject of many backyard scientific studies, and this process fascinates even those who have witnessed it many times over. Through the course of a week or two, legs grow, the tail shrinks, the gills recede, lungs are formed and a whole manner of more subtle changes mold an aquatic animal into a more or less terrestrial one. This outstanding transformation is surprisingly domestic and takes place in ponds and ditches near to almost every Canadian.

Best Sites Map
Canada's National Parks in green.
Provincial Parks in blue.

There's no place like home

To no one's surprise, large areas of Canada are not populated by either amphibians or reptiles. Although a few species have the ability to survive eight months of winter, conditions above the treeline—whether on mountains or in the Arctic—are simply too harsh for these animals to take hold. Conversely, amphibians and especially reptiles are found in higher numbers and in greatest diversity in the warmest regions of the country.

The ***temperate West Coast*** of British Columbia features Canada's mildest and wettest region. A unique assemblage of species occurs here, principally composed of West Coast species that reach their northernmost distribution in B.C. The ***dry, interior Okanagan*** and other valleys of southern B.C. are also of interest. Here, in prairie-like conditions, are species that are different from coastal forms and different, too, from animals living in a similar situation on the east side of the Rockies.

The ***grasslands*** of Alberta, Saskatchewan and Manitoba feature hot, dry summers and cold winters. These conditions restrict many species typical of desert systems to their northernmost limit. Although the number of species is low in comparison to some regions, the grasslands region of Canada has many unique species.

The ***Great Lakes region*** is the motherlode of reptiles and amphibians in Canada. South of a line arcing from Windsor through Ottawa to Quebec City lies the most diverse representation of reptiles and amphibians north of the United States. Eastern North America features a high diversity of reptiles and amphibians—amphibians in particular. This high diversity in the U.S. seeps north across the border into the heavily populated regions of southeastern Ontario and southern Quebec. Not surprisingly, the decline of reptile and amphibian populations in these regions is a direct result of an increase in human population and activity.

The ***Maritime region*** has its share of the eastern reptile and amphibian wealth, but many species do not continue uninterrupted into central Canada. For the most part, reptiles and amphibians have the heart of their distribution in the Appalachians, and it's from there that Canadian populations are presumed to arise. For this reason and because it is relatively warm, the southern third of Nova Scotia is particularly productive for reptile and amphibian diversity. Long ago, this peninsula was connected to the United States by a land bridge. This bridge, now cut off by the Bay of Fundy, allowed animals to populate regions that now are discontinuous with their original habitat. As a result, we see populations of reptiles and amphibians in Kejimkujik National Park and in the Annapolis Valley that are now isolated from populations of the same species to the south and to the west.

The Good, the Bad and the Misconceptions

For too many people, fear and disgust are the dominant emotions when meeting a reptile or amphibian. Snakes, of course, bear the brunt of

a disproportionate share of ill will. The source of "snake scare" is not fully understood, but it is clear that these animals have been typecast as the bad guys for quite some time. From the story of Genesis in the Bible to the children's classic *The Jungle Book*, snakes have long been demonized. Their unemotional gaze, venomous reputation and slithering form all contribute to this stereotype. Of course, if one encounters a venomous snake, it makes sense to show a healthy respect for it. Indeed, it is likely that some of our fear of snakes arises from an evolved response to venomous species. However, in Canada we have only three, mildly venomous species, and, undoubtedly, much of our fear is learned, exaggerated and fueled by misconceptions.

Snakes are not alone in producing fear and mistrust in some people. Toads also suffer. Unfortunately, the false connection between touching a toad and getting warts continues to be kept alive in spite of all reasonable evidence to the contrary. Although these misgivings seem superficially harmless, many individual snakes, toads, snapping turtles and other reptiles and amphibians are needlessly killed out of fear and ignorance.

Fortunately, some amphibians and reptiles benefit from good publicity. Frogs and turtles, which are featured in popular children's stories and are portrayed by such loved characters as Kermit the Frog and the Ninja Turtles, do not tend to suffer from negative stereotypes and even have positive reputations. People communicate at some level with turtles because they can read emotions of fear and curiosity with each tucking and untucking of their heads. Frogs may have an even stronger connection with people, as their large, often brightly coloured eyes, attractive colour patterns and obvious harmlessness make them familiar and worthy of our empathy.

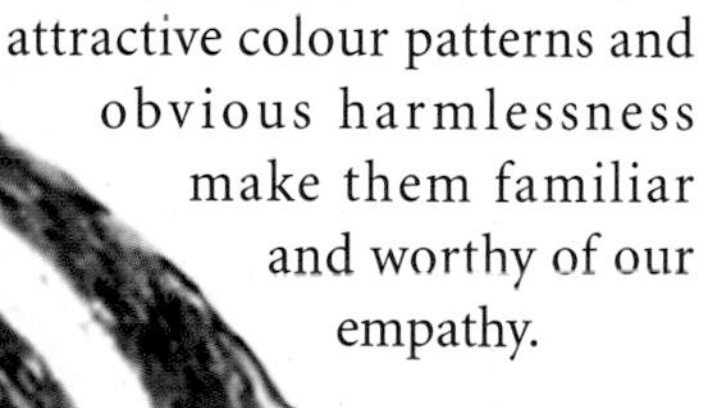

Hands On

For the most part, amphibians and reptiles in Canada are completely harmless. Rattlesnakes are the most obvious exception to the rule, but despite all their bad publicity, conflicts between rattlesnakes and humans are extremely rare. Certain toads, frogs and salamanders, if ingested (or licked), would lead to some intestinal hardship, and some lizards and snakes like to bite. But as interactions most often come on our terms and not theirs, it is understandable that a wild animal will defend itself if mistreated. In fact, one of the most wonderful characteristics of reptiles and amphibians is that they provide people an intimate relationship with a wild animal.

Unlike birds that are distant and mammals that are either too large or too fast, reptiles and amphibians can be caught and admired in the hand. Rich and powerful bonds to the natural world are easily formed in such encounters—witness the annual pilgrimages of children to frog ponds worldwide. The unique ability to hold a wild animal and to know it as an individual strikes strong emotional chords within us all.

There is really nothing wrong with handling reptiles and amphibians, provided that you follow a code of respect. All animals should be returned to their point of capture, and there should be minimal disturbance to their habitat during a search. Because many of these animals are found beneath boards, rocks or logs, you must always roll the object toward you rather than away. Doing so ensures an escape route for both you and an animal should any unanticipated surprise be uncovered. When returning an animal to its home, replace the cover first and then encourage the animal to return to safety.

Physically handling reptiles and amphibians holds inherent risks to the small animals. Snakes should always be completely supported; bones can easily break if an animal is picked up by one end. Use both hands to support the entire weight of the animal. Do not handle venomous or agitated snakes at all. Large frogs and toads should be restrained by grasping their thighs with one of your hands and supporting the body with the other. Small frogs and salamanders are best observed through a water-filled, clear plastic bag. Immersion in water prevents these animals from drying out, reduces their stress level and allows them to be safely passed from person to person for better observation. Always wet your hands and make sure that any lotion or bug spray has been washed off before touching an amphibian. Their skin can absorb compounds straight from your skin.

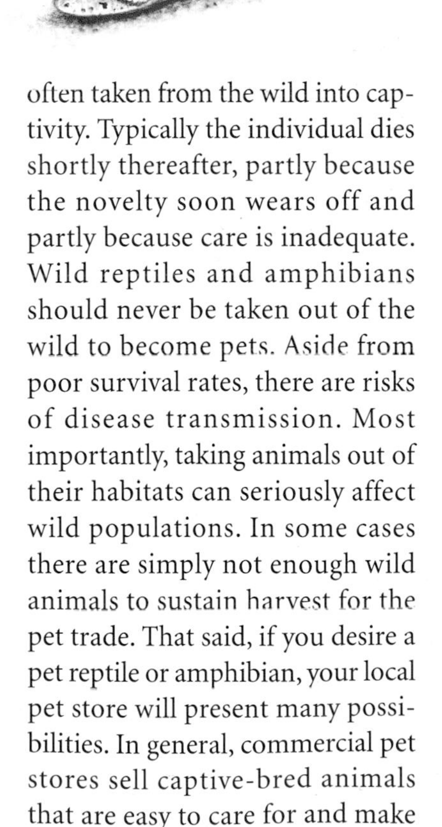

It is very natural to establish personal bonds during these intimate encounters with reptiles and amphibians. Breaking these bonds can be quite difficult, particularly for children. For this reason, individual reptiles and amphibians are often taken from the wild into captivity. Typically the individual dies shortly thereafter, partly because the novelty soon wears off and partly because care is inadequate. Wild reptiles and amphibians should never be taken out of the wild to become pets. Aside from poor survival rates, there are risks of disease transmission. Most importantly, taking animals out of their habitats can seriously affect wild populations. In some cases there are simply not enough wild animals to sustain harvest for the pet trade. That said, if you desire a pet reptile or amphibian, your local pet store will present many possibilities. In general, commercial pet stores sell captive-bred animals that are easy to care for and make excellent pets. Organizations such as the Canadian Amphibian and Reptile Conservation Network (CARCNET) and the Kawartha Turtle Trauma Centre (KTTC) can provide more information on pets and on captive rehabilitation.

It Ain't Easy Being Green

In this era of environmental sensitivities, an increase in awareness of a group of animals must also include their conservation plight. Some reptile and amphibian populations are expanding, but as a whole reptiles and amphibians are declining at a sharper rate than any other group of species. One-third of the species described in this book, including two-thirds of Canada's reptiles, have been designated by the Committee on the Status of Endangered Wildlife in Canada (COSEWIC) as being at risk in varying degrees: Endangered, Threatened or Special Concern. This status is particularly alarming given the evolutionary resilience that these animals have demonstrated throughout their existence.

Amphibians in particular are highly sensitive to environmental conditions. Through both aquatic and terrestrial life stages, amphibians are exposed to many more environmental factors than species that are solely terrestrial or aquatic. Their sensitivity is compounded further by porous skin that does not shield against environmental contaminants. Increased ultraviolet radiation,

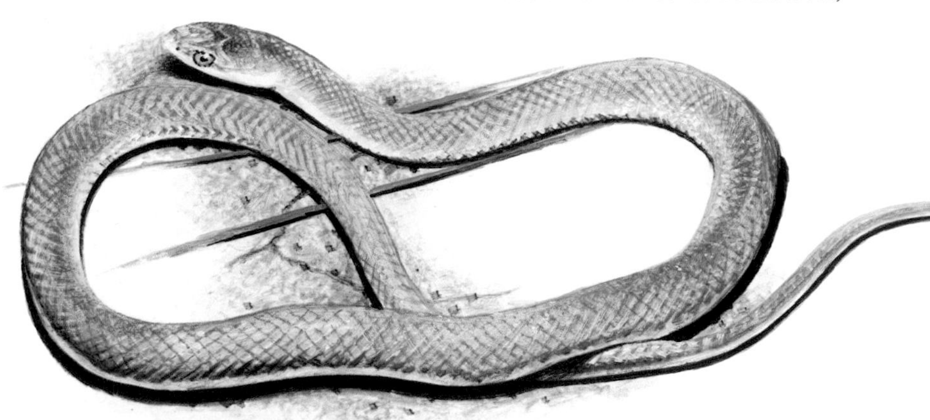

pesticides, climate change, drought, salination, introduced species and habitat destruction have all been implicated in amphibian declines in recent years. These animals are barometers of environmental health, and when populations of frogs, toads and salamanders begin to disappear at accelerated rates, it's time to listen to the message.

The Species Accounts

This book provides accounts of 84 reptiles and amphibians found in Canada today. All of these descriptions involve reproducing populations. In addition, there is a section at the end of the Reptiles portion of the book containing mini accounts of several other reptile species associated with Canada. The seaturtles included do not breed in Canada, but they do make periodic migrations into Canadian waters. The introduced species now live in, but are not native to, Canada. And the extirpated species are those that can no longer be found in Canada, though they may continue to survive elsewhere.

Much of the species account is devoted to a unique portrayal of the animal's character. A deliberate attempt has been made to celebrate the experience of meeting these animals in the wild on their own terms. It is hoped that the animal's spirit is shared, even with readers who may never have had a first-hand experience with reptiles and amphibians. The species are described as one would introduce a friend, to reflect admiration and to make readers want to get to know and care for them.

The remainder of the individual account offers more traditional information. Because the philosophical approach of this book is to inspire field experiences, the **ID section** is of great value. Field marks are given in order of importance, with easily observable traits stressed. Some species are not easy to differentiate from one another without close scrutiny, but even in these situations care has been taken to limit confusion.

The **Length** of all animals is given as total length, from snout to tail tip. For frogs and toads, the length of only the body is given, not the body plus the length of outstretched legs.

The **Distribution** of the species is of great benefit to the identification. To encourage you to discover the wonderful world of reptiles and amphibians and to streamline your search for a particular animal, one or several **Selected Sites** for the species have been included. For a list of abbreviations, see page 204.

Within its range, each reptile or amphibian has a preferred **Habitat**. Do not be completely limited by the habitat described herein; in rare instances, individuals of any species can be found in the oddest of places.

Activity Patterns explains the daily and seasonal movements of each reptile and amphibian. One seasonal movement is often to breeding areas. The **Reproduction** of these animals is always interesting because it differs between species. Here you will find breeding behaviour, number of offspring and the time needed for each process involved.

What these animals eat and the differences in the diet of their different life stages are listed in the **Food** section.

When appropriate, the **Call** of particular amphibians is described in as accurate a manner as possible using the written word. We suggest that if you are interested in meeting frogs and toads in particular, you should learn their voices. Excellent recordings are available for species in your area on CDs and at various web sites.

A brief list of **Similar Species** and their differing characteristics is provided to complement the ID section. By narrowing down the list of possible suspects, your process of identification is made easier.

The **French Name** of each species is given as a bit of interesting information because we are Canadian. And last, because people really like to embrace and share trivia, there's a **Did You Know?** section for those of you in search of the perfect fact to astound your friends with.

Reptiles

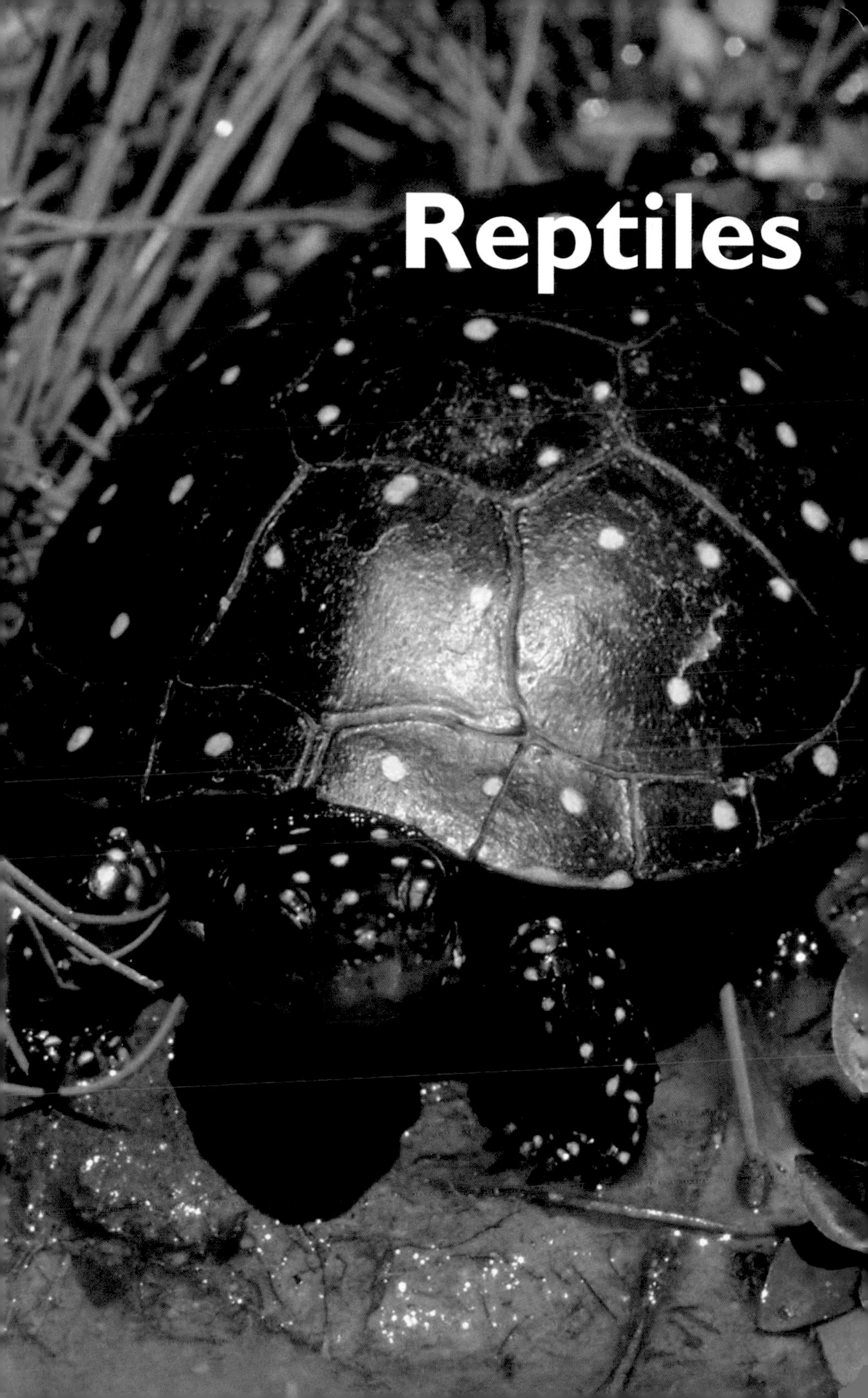

Snapping Turtle

Chelydra serpentina

Legends abound about Snapping Turtle encounters, but this calm reptile is usually mired in the muddy reaches of its favourite waterways with only its eyes and nostrils exposed. Only the movement of a potential meal rouses a Snapping Turtle at rest in its soft mud bath. • A Snapping Turtle that is accidentally stepped on while lying in the water rarely strikes back, opting instead to sheepishly tuck in its head in response. Its shy, almost hermit-like manner means that usually only the most dedicated herpetologists encounter "Snappers." • The Snapping Turtle's sturdy legs are capable of satisfying any prehistoric urges for overland adventure. Pregnant females are the most commonly encountered on land, searching for suitable soil in which to dig a nest and lay eggs. In contrast to a Snapper's behaviour in water, in terrestrial settings it is defensive, willing to strike out at anything that violates its safety zone. However, accounts of excised toes are exaggerated: the sharp, strong beak of this animal is capable of breaking the skin but not much else. • Really, the Snapper deserves a little sympathy. Its quest for suitable nesting grounds often meets with tragedy while crossing roads.

SIMILAR SPECIES:

Painted Turtle, p. 32

Wood Turtle, p. 38

DID YOU KNOW? Never grasp the animal's tail to pick it up, as this can easily dislocate vertebrae and damage the spinal cord. If you must pick up a Snapper, grasp it with both hands by the rear of the upper shell.

It is a noble thing to encourage any Snapping Turtle you find on the shoulder of a road to a safer location. Just be sure which direction the turtle was walking, or you could unintentionally erase an entire day's effort! And remember to watch for cars yourself—you don't want to be roadkill any more than the turtle does.

ID: largest freshwater turtle in Canada; 3 rows of knobs along top of shell (carapace); shell is dark green to black, but shell on older turtles can be covered by dark algal growth; breastplate (plastron) is light tan or grey; saw-tooth pattern marks a uniquely long tail.

LENGTH: *Male:* 25–47 cm (10–18½"). *Female:* 22–35 cm (8½–14").

DISTRIBUTION: from Saskatchewan east to the Maritime provinces. *Selected Sites:* Kejimkujik NP (NS), Algonquin PP (ON).

HABITAT: freshwater wetlands; stagnant, mud-bottomed wetlands and thick vegetation; occasionally seen on land during dispersal or during search for nest sites.

ACTIVITY PATTERNS: active from mid-spring to fall; hibernates all winter on the bottom of lakes and rivers.

REPRODUCTION: mating can occur at anytime in the year that the turtles are active; they do not mate until they are at least 15 years of age; in late spring females seek out an area with loose soil in which to deposit their eggs; average females will lay 25–50 eggs that hatch during fall, but extraordinary clutches can produce over 100 eggs; sex is dependent on incubation temperature; warm or cool temperatures produce females, and intermediate temperatures produce males.

FOOD: omnivorous; plants, aquatic invertebrates, fishes, frogs, toads, snakes, small turtles, aquatic birds and carrion.

SIMILAR SPECIES: *Painted Turtle* (p. 32): smaller; more colourful face and legs; smooth shell; short tail. *Wood Turtle* (p. 38): smaller; shell has light lines radiating from ridges.

FRENCH NAME: Chélydre serpentine

Stinkpot
Sternotherus odoratus

Plenty of great things can be found in southern Ontario, but vast wilderness is not one of them. Fortunately for herpetologists and Stinkpots, there are a few wetlands that remain for these animals. • Stinkpots are small and inconspicuous among the turtle crowd in southern Ontario. They are difficult to find because of their elusive nighttime venturing and their sub-aquatic preferences. When they are seen, you can often rely on that location holding on to its turtles for some time. Stinkpots are homebodies, and they prefer not to venture far—many individuals will remain in their natal pond for all of their (usually) 25 years. • Stinkpots are classic neck-tuckers, withdrawing into their shells at the meekest stimulation. The shell does not tighten up as snugly as some, so even when fully withdrawn, the animal's eyes and small smile are seen tucked deep within the shell. Legs and tail also withdraw, but even less effectively than the head. For this reason, older Stinkpots are often found without toes, legs or tail. Being chiefly aquatic, these amputees still manage to move through the water with a reasonable degree of grace. • Stinkpots are very effective residents of the underwater world. Turtles are different from fishes and crustaceans that breathe through gills, and turtle skin is not as permeable to dissolved oxygen as that of amphibians. But, while underwater, these turtles use only one-eighth of the oxygen they require when breathing air on land. This means that they can remain submerged for up to 20 minutes, although typical dives are much shorter.

SIMILAR SPECIES:

Painted Turtle, p. 32

Blanding's Turtle, p. 36

Spotted Turtle, p. 34

ID: 2 light stripes surround eye and continue down neck; highly domed shell; underparts are yellowish brown; 1 hinge. *Juvenile:* prominent keel along back; scattered spots or streaks; grey-brown to black.

LENGTH: 7–13 cm (3–5").

DISTRIBUTION: patchy populations in southern Ontario and southern Quebec; they have largely disappeared from southwestern Ontario south of the Canadian Shield. *Selected Sites:* Georgian Bay (ON).

HABITAT: shallow lakes, rivers and wetlands with muddy bottoms.

ACTIVITY PATTERNS: primarily active after dark; active when water temperatures rise above 10° C (50° F); often bask on the water surface beneath lily pads; below 10° C they will burrow into muddy bottoms and remain inactive.

REPRODUCTION: nesting occurs from May through July; nests are on land but close to water; nests may be on open ground or under litter; some females may lay their eggs in the same nest; 2–5 eggs are laid; eggs hatch in fall; incubation temperature determines sex of hatchlings; males reach maturity at 3–4 years old, but females do not reach maturity until 9–11 years old.

FOOD: omnivorous; small invertebrates, amphibians and small fishes, as well as algae.

SIMILAR SPECIES: *Painted Turtle* (p. 32): more yellow or red markings on shell and legs. *Blanding's Turtle* (p. 36) and *Spotted Turtle* (p. 34): more distinct light spots or streaks on shell; shell is much broader.

FRENCH NAME: Tortue musquée

Spiny Softshell

Apalone spinifera

This may be the easiest of the turtles in Ontario to identify, even though it is rare and very seldom seen by anyone not staring intently into just the right pond. The smooth, sleek Spiny Softshell looks a bit like a giant wet olive, and both the carapace and the plastron are soft, smooth and rubbery rather than bony and horny like the shells of most other turtles.

- Contrary to its cute exterior is its cantankerous attitude. Any handling of a wild Spiny Softshell is an almost certain guarantee to be bitten and scratched. With its long neck, sharp beak and quick movements, a Softshell can find a finger to bite in places that are safe to hold on most other turtles. This animal is fast, aggressive and flexible because of its lack of a hard, protective shell, and there is no good reason to handle one just for fun.
- The Spiny Softshell's long nose and neck act as a built-in snorkel, enabling it to breathe safely while completely submerged. Although breathing air is its most direct means of acquiring oxygen, almost half of its needs can also be met through the absorption of dissolved oxygen through its soft

SIMILAR SPECIES

Stinkpot, p. 28

skin and body orifices. • Sanctuary for this animal lies below the water's surface; few animals are able to kill an adult Softshell in this realm. A young turtle, however, is vulnerable to predators such as raccoons, skunks, large fishes and wading birds. Humans remain this turtle's largest threat, as habitat destruction has removed most of its nesting areas. When forced to move from altered wetlands, the Spiny Softshell is even less successful than other turtles. Away from water, it rapidly becomes desiccated and dies. Fortunately, many of the remaining areas of southern Ontario that continue to boast populations of this Canadian turtle have been afforded some level of protection.

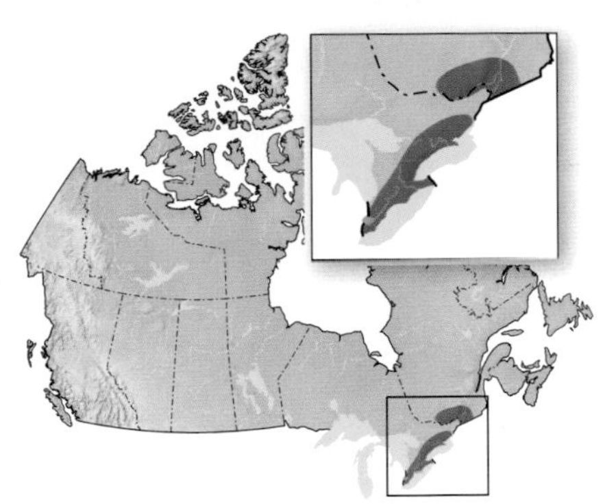

ID: flexible, leathery, olive-coloured shell lacks scutes and has small, dark spots with black outlines that are broken in females and solid in males; long, tubular snout; 2 dark-bordered, light stripes on grey to olive head; white or yellow underside.

LENGTH: *Male:* 12–23 cm (5–9"). *Female:* 18–42 cm (7–16½").

DISTRIBUTION: extreme southwestern Quebec and around Lake Erie and Lake Ontario. *Selected Sites:* Rondeau PP, Long Point PP, Thames River (near Fanshawe Dam), Sydenham River (all in ON).

HABITAT: almost always found in or very near water; rivers or lakes with soft, muddy bottoms and heavy aquatic vegetation.

ACTIVITY PATTERNS: active during the day, generally during the warmest periods from late spring to early fall.

REPRODUCTION: nesting occurs in June and July in sandy soil; 1 or possibly 2 clutches of 12–40 eggs are laid each year; incubation temperature of the eggs does not determine sex; eggs hatch in August or September; sexual maturity is likely not attained in females for more than 10 years in northern populations.

FOOD: insects, crayfish and other invertebrates; occasionally small fishes, amphibians and aquatic plants.

SIMILAR SPECIES: not likely to be confused with any other turtle. *Stinkpot* (p. 28): smaller; shell has scutes.

FRENCH NAME: Tortue-molle à épines

DID YOU KNOW? The Spiny Softshell is not all that spiny in comparison to other reptiles. Small spines are found along the leading edge of the shell, but they are not obvious field marks. This turtle is listed as Threatened by COSEWIC.

Painted Turtle

Chrysemys picta

Western subspecies

The Painted Turtle is the most widespread turtle in Canada as well as the United States. Painted Turtles have a habit of basking in conspicuous places just out of the water, allowing for easy observation. On a premium basking site such as a partially submerged log along the edge of a wetland, Painted Turtles can be lined up like concert-goers awaiting a big ticket purchase. Some sleep, others are vigilant while still others clamour over neighbours, hoping to find the sweetest sun spot. These turtles are wary, and when a strange presence approaches they quickly enter the water, flopping from their basking site like a collapsing house of cards. • All this basking serves an important purpose, as turtles have little ability to generate internal heat. For Painted Turtles to be operational, their internal temperature needs to be over 20° C (70° F) and the water must be at least 15° C (60° F). During winter, the turtles bury themselves in the soft mud of lake bottoms. The unique physical property of water has the liquid at its most dense (and therefore heaviest) at 4° C (40° F). Therefore, even if the lake is ice covered, the bottom remains a relatively cozy 4° C, which enables the turtles to survive long periods without having to refuel their slowed metabolism.

ID: yellow stripes on head and neck; red on legs and underside margins of shell; base colour ranges from olive to black; flat, smooth shell. Three subspecies are found in Canada. *Eastern* subspecies: Maritime provinces; plain yellow plastron. *Midland* subspecies: eastern Ontario and Quebec; yellow plastron with a small, central, dark pattern; poorly defined mid-dorsal

SIMILAR SPECIES

Northern Map Turtle, p. 40 Red-eared Slider p. 105

DID YOU KNOW? Painted Turtles in the northern part of their range may live to be over 100 years old. The western subspecies is listed as Endangered in southwest British Columbia and Vancouver Island and as Special Concern in the interior of B.C. by COSEWIC.

stripe on carapace. *Western* subspecies: Lake Superior to the West Coast; largest subspecies; light, wavy lines on shell; complex yellow and red markings all over plastron.

LENGTH: 10–25 cm (4–10").

DISTRIBUTION: southern Canada from British Columbia to Nova Scotia; fewer on the prairies of Alberta and Saskatchewan and declining in B.C.; have been introduced to areas outside of their natural range. *Selected Sites:* anywhere within their range and habitat.

HABITAT: ponds, lakes, oxbows and slow creeks and rivers with moderate amounts of aquatic vegetation; move to and from deep wintering wetlands.

ACTIVITY PATTERNS: become active in mid-spring when water reaches 15° C (60° F); active during the day; hibernate buried in the muddy bottoms of lakes and large ponds during winter.

REPRODUCTION: female digs a shallow (often less than 12 cm [5"] deep) nest from late May to early July in sandy or loose soil; 5–20 eggs are laid; female may lay a second clutch; young turtles hatch and may emerge in fall but usually overwinter in the nest and emerge the next spring; hatchlings can tolerate being frozen to as low as -11° C (12° F); eggs incubated at temperatures above 30° C (85° F) produce all females, lower temperatures produce more male offspring; males sexually mature at 8 years, females at 12–15 years.

SIMILAR SPECIES: *Northern Map Turtle* (p. 40): not as colourful; end of shell is notched. *Red-eared Slider* (p. 105): red "ear" band streaking back from eye.

FRENCH NAME: Tortue peinte

Spotted Turtle

Clemmys guttata

This CD-sized turtle is not often seen in Canada, the northern limit of its North American range. The Spotted Turtle is small, has a limited distribution and has secretive tendencies. Movements overland happen under the cover of rain. But the Spotted Turtle does enjoy a good soaking of sun, and it is at basking sites that it is most frequently observed. • This turtle is a cool weather reptile and is active very early in spring. During the warmest period of southern Ontario summers, it retreats to cool soil or lies beneath a quilt of dead leaves and enters a quiescent state of dormancy. Although it would be unfair to characterize this animal as lazy, a certain level of inactivity is required for it to withstand conditions here in Canada at its northern limit.

ID: small and dark with round, yellow or sometimes orange spots on black carapace, head, neck and limbs; domed, smooth shell; creamy yellow plastron with large, black markings on each scute unless entirely black; shell lacks a hinge. *Male:* dark jaw. *Female:* yellowish jaw.

LENGTH: 9–14 cm ($3^1/_2$–$5^1/_2$").

DISTRIBUTION: southern Ontario and southeastern Quebec. *Selected Sites:* Georgian Bay region, Rondeau PP (both in ON).

HABITAT: large ponds and marshes and small, muddy-bottomed lakes.

SIMILAR SPECIES

Blanding's Turtle, p. 36

Stinkpot, p. 28

Wood Turtle, p. 38

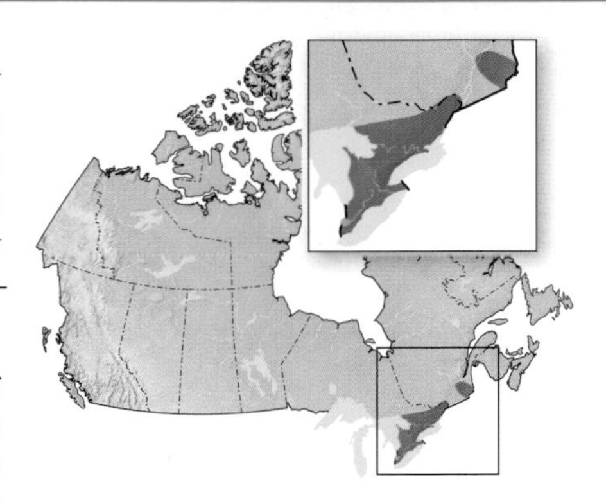

ACTIVITY PATTERNS: generally active during the day; emerges from winter hibernation in early spring and typically enters dormancy once again in early fall; if too hot in summer, may enter a short-lived period of dormancy.

REPRODUCTION: mating can occur anytime the turtles are active; female can store sperm until ready to fertilize eggs; 3–7 eggs are laid in June; nest is dug in sandy soil in an open, sunny area; sex is temperature dependent; eggs hatch in early fall; sexual maturity is often reached by 12 years.

FOOD: snails, aquatic insects and aquatic vegetation.

SIMILAR SPECIES: *Blanding's Turtle* (p. 36): bright yellow chin and throat; highly domed shell; more irregular markings. *Stinkpot* (p. 28): much narrower shell that is not marked when it is an adult. *Wood Turtle* (p. 38): lacks yellow spots on shell; rougher shell.

FRENCH NAME: Tortue ponctuée

Blanding's Turtle

Emydoidea blandingii

With little more than their heads emerging from a weedy wetland, many turtles are difficult to identify. Not so for the Blanding's Turtle—its chin and throat are such a vivid shade of yellow that this turtle's true identity is betrayed. The Blanding's Turtle is a favourite among Canadian turtle watchers; it is not only easy to identify, but this attractive northern specialist often basks conspicuously on floating logs. Aside from its trademark chin, its shell is often adorned with light spots and streaks. It is thought that these markings are fashioned by evolution to mimic the patterning of duckweed, a pervasive floating aquatic plant that the Blanding's Turtle is often found among. • Like the Stinkpot and the Eastern Box Turtle, the Blanding's Turtle has a hinge midway on its underside. This ligament hinge allows

SIMILAR SPECIES

Northern Map Turtle, p. 40

Spotted Turtle, p. 34

the underside to pull up, sealing the turtle up very tightly when it chooses to retract, protecting it from danger. Even with this protective feature, typically about a third of the individuals in a given population show injuries from encounters with predators such as raccoons. Chewed shells and missing tails and toes all attest to the natural dangers that cause this turtle concern.

ID: luminescent yellow chin and throat; long neck; mouth curves up in a "smile"; dark, rounded carapace has irregular, light markings that may be faded in older individuals; underside is hinged and can be yellow with blotches or sometimes entirely black; big, protruding eyes.

LENGTH: 18–26 cm (7–10").

DISTRIBUTION: southern Ontario and southern Quebec; an unconnected population exists in Nova Scotia around the Annapolis Valley and Kejimkujik NP. *Selected Sites:* Big Creek Marsh at Long Point (ON), Ottawa Valley (ON and QC), Kejimkujik NP (NS).

HABITAT: vegetated wetlands, including lakes and ponds with muddy and organically rich bottoms.

ACTIVITY PATTERNS: active during the day, particularly in the morning hours; most frequently basks in spring; active from mid-April to September.

REPRODUCTION: has exceptionally late maturity; females do not reach sexual maturity until they are 15–25 years old; up to 15 eggs are laid in a nest dug in sandy soil in mid- to late June; sex of offspring is temperature dependent; young will spend much of their time amongst thick wetland vegetation.

FOOD: mainly crayfish; also insects, fishes, frogs and plants.

SIMILAR SPECIES: no other Canadian turtle has a bright yellow chin and throat. *Northern Map Turtle* (p. 40): yellow lines on neck and legs; rear of shell is saw-like. *Spotted Turtle* (p. 34): much smaller; distinct yellow spots throughout body.

FRENCH NAME: Tortue mouchetée

DID YOU KNOW? The dome shape of its shell indicates that the Blanding's Turtle is not a great swimmer—it mostly moves by walking along the bottom of its pond. Unlike many aquatic turtles, the Blanding's Turtle has the ability to swallow food on land. The Blanding's Turtle is listed as Threatened in the Great Lakes area and as Endangered in Nova Scotia by COSEWIC.

Wood Turtle

Glyptemys insculpta

Algonquin Provincial Park is one of Canada's best sites for a mini tortoise known as the Wood Turtle. Despite its name, the Wood Turtle is as strongly associated with water as it is with woodlands. Wood Turtles move very well over land, particularly following heavy summer rains that spawn the growth of mushrooms or berries—favourite meals of the Wood Turtle. • This turtle has a wise look to its wrinkled face, which just seems to reinforce observations by scientists and pet owners alike that it is indeed rather smart. Unfortunately, as a result of its endearing nature, many populations in the United States and a few in Canada have in the past been decimated by the collection of individuals for the pet trade. And although the Wood Turtle's reputed peak sprinting speed of 0.32 kilometres (0.20 miles) per hour places it in gold medal standing among Canada's turtle contingent, it is not fast enough to outrun even the slowest turtle collectors. Collectors,

SIMILAR SPECIES

Blanding's Turtle, p. 36

Spotted Turtle, p. 34

coupled with habitat degradation, have reduced the Wood Turtle's Canadian range. Collection is now illegal for all native turtle species, and also frowned upon by all those who care about Canada's wildlife as turtle populations cannot withstand even low levels of collection.

ID: highly domed shell with raised pyramidal ridges giving a sculpted appearance; shell is light to dark brown with light lines radiating from each ridge; underside is yellow with dark blotches; underside is not hinged; head is dark and may have pale yellow spots; throat, underside of tail and legs are yellow to red; plastron is concave in adult males.

LENGTH: 16–25 cm (6–10").

DISTRIBUTION: from southern Ontario and Quebec across to New Brunswick and Nova Scotia. *Selected Sites:* Algonquin PP (ON).

HABITAT: forages in upland forests and meadows; hibernates in water.

ACTIVITY PATTERNS: active from mid-spring to mid-fall; active mostly during the day but occasionally on warm (over 20° C [70° F]) nights.

REPRODUCTION: mating occurs in late spring and fall; nesting occurs in June; females dig a nest in a sandbar or in other loose soil near water then deposit up to 20 eggs; sex is independent of temperature; sexual maturity is usually attained by 18 years of age.

FOOD: vegetation, fruits, flowers, mushrooms and invertebrates.

SIMILAR SPECIES: *Blanding's Turtle* (p. 36): bright yellow throat; smoother carapace; light-coloured spots on carapace. *Spotted Turtle* (p. 34): smaller; carapace is black, spotted and smoother.

FRENCH NAME: Tortue des bois

DID YOU KNOW?
The Wood Turtle has been observed "worm stomping." Typically, an individual stomps its feet and shell on the ground to encourage earthworms to come to the surface, where the turtle catches and eats them. The Wood Turtle is listed as Special Concern by COSEWIC and as Endangered in Ontario by COSSARO.

Northern Map Turtle

Graptemys geographica

Northern Map Turtles received their name based on the yellow lines that crisscross the backs of young individuals, resembling the waterways on a map. However, the lines do fade over time. • These shy turtles like to stay away from the hubbub of the wetland rim, remaining in deeper water. They prefer the crunchier snacks; their favorite foods include crayfish, snails and mollusks. • Northern Map Turtles are a hardy species, proving their worthiness as Canadian turtles by being active under the ice, crawling along pond bottoms in search of some winter treats. This species can be one of the first spring sightings and one of the last fall views as it finally retreats into the water to wait out winter. During late spring, look for groups of Northern Map Turtles

SIMILAR SPECIES

Blanding's Turtle, p. 36

Painted Turtle, p. 32

basking on a log out in the pond—many of the pregnant females will warm up their bodies to help speed up the development process.

ID: hexagonally divided shell with ridges along the bottom and along the vertebral line; heavy, yellow markings against a dark background on head and legs. *Male:* longer tail.

LENGTH: *Male:* 10–16 cm (4–6"). *Female:* 17–27 cm (6½–10½").

DISTRIBUTION: southern Ontario and southeastern Quebec; can be found along large rivers like the St. Lawrence. *Selected Sites:* Ottawa River (ON and QC), Rondeau PP, Pelee Island, Point Pelee NP (all in ON).

HABITAT: prefer slow moving, large rivers or reservoirs with pools and debris; require habitats with clean, deep water; can also be found in ponds and lakes.

ACTIVITY PATTERNS: active during the day; may be active during warmer winters.

REPRODUCTION: breed after emerging in spring; females nest in June and July in sandy areas; 12–20 hatchlings emerge in fall; sex of hatchlings is temperature dependent during incubation; males mature a few years earlier than females; females likely mature at 10–14 years, though this is not known for sure for Canadian populations.

FOOD: primarily mollusks, crayfish, and snails; also aquatic plants, insects and fishes opportunistically.

SIMILAR SPECIES: *Blanding's Turtle* (p. 36): bright yellow chin; smoother shell; no defined markings on head and legs; shell lines wavy and not as well-defined. *Painted Turtle* (p. 32): smoother shell; red markings along edge of shell, sometimes on legs and neck; red on plastron.

FRENCH NAME: Tortue géographique du nord

DID YOU KNOW? Molluscs, an important food source for Northern Map Turtles, require clean water, something that is getting harder to find throughout their Canadian range. Northern Map Turtles are listed as Special Concern by COSEWIC.

Greater Short-horned Lizard

Phrynosoma hernandesi

Greater Short-horned Lizards masterfully combine cryptic camouflage with a reclusive lifestyle to evade detection. Should you find yourself in the extreme southern part of Alberta or Saskatchewan, do not presume that you will find one. Many dedicated naturalists have spent hours staring hypnotically at the loose rocks of the badlands without spotting one of these elusive reptiles. With camouflage and seclusion already limiting human encounters, Greater Short-horned Lizards complete the hat trick of viewing obstacles by remaining inactive under all but the most perfect weather conditions. Only when the sun emerges are Greater Short-horned Lizards apt to be coaxed from their hiding spots to pursue their ant-eating ways. Under these conditions the stealthy reptiles roam the rocky landscape, travelling in well-spaced bursts of speed. • Greater Short-horned Lizards are so uncommon that they have no specialized predators and need only worry about

DID YOU KNOW? Greater Short-horned Lizards are quite famous for being able to shoot blood from their eyes, a startling reaction to the stress of an encounter with a predator. In reality this feat is far less dramatic than it seems, and it is very seldom encountered because these lizards don't have many predators. Greater Short-horned Lizards are listed as Special Concern by COSEWIC.

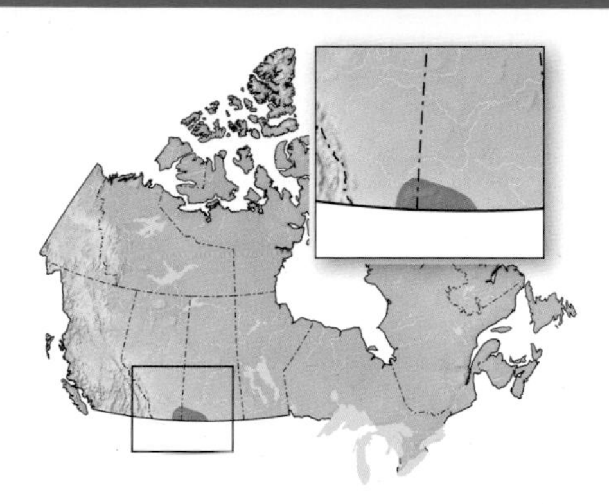

opportunistic animals such as coyotes. Rather than running away at the first signs of danger, these miniature lizards rely on their camouflage to keep them out of sight of potential predators and will stay still when spotted. From a naturalist's perspective, this defence strategy can be a tremendous benefit. Upon finally locating a Greater Short-horned Lizard, you will be pleasantly surprised by its willingness to pose for photographs...a nice reward for all those years of searching.

ID: grey, yellowish or reddish brown mottled camouflage all over body; small "horns" on head.

LENGTH: *Male:* 5 cm (2"). *Female:* 7–11 cm (3–4½").

DISTRIBUTION: southeastern Alberta and southwestern Saskatchewan. *Selected Sites:* Grasslands NP (SK), Milk River region (AB), badlands around Manyberries (AB).

HABITAT: badlands country; loose shale or sun-baked rocky outcrops; fine clay bottomlands (watch where you step!).

ACTIVITY PATTERNS: typically active only on the hottest days in the hottest months of the year; during cooler periods and winter, they burrow into gravelly, sandy areas to avoid the cold.

REPRODUCTION: give birth to live young; up to 13 young, about 2.5 cm (1") long, are born in July or August.

FOOD: primarily ants; diet is probably supplemented with other terrestrial invertebrates.

SIMILAR SPECIES: the only naturally occurring lizard in Alberta and Saskatchewan.

FRENCH NAME: Iguane à petites cornes

Northern Alligator Lizard

Elgaria coerulea

The only alligators in Canada are lizards, and they would fit in your hand. Although British Columbia's lizard may share a common look with the tropical giants, and despite its size is a pretty tough animal, the Northern Alligator Lizard has very little in common with the southern swamp dwellers. Our lizard prefers dry land to water. Although a conventional alligator wouldn't last one winter in Canada, the Northern Alligator Lizard thrives throughout much of southern British Columbia. • Understandably, Canada does not get a high ranking when it comes to lizard diversity—our paltry five species total does not compare to the world's 4500—but the Northern Alligator Lizard represents its kind well in our country. It is a skinny, fast-footed lizard with a passion for sun bathing. Although it scurries away at the first sign of danger, it usually returns faithfully to its

SIMILAR SPECIES

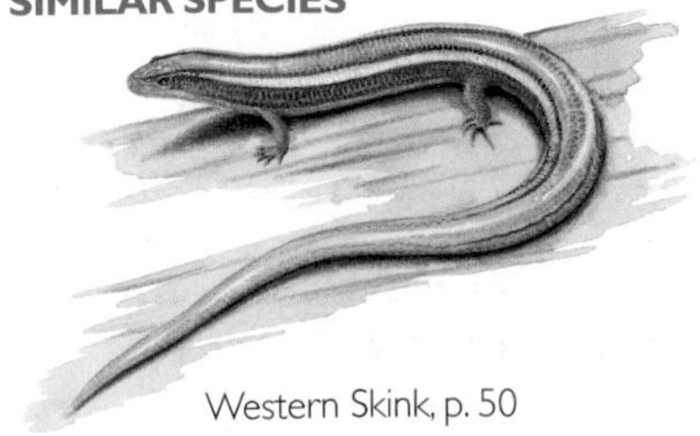

Western Skink, p. 50

European Wall Lizard, p. 106

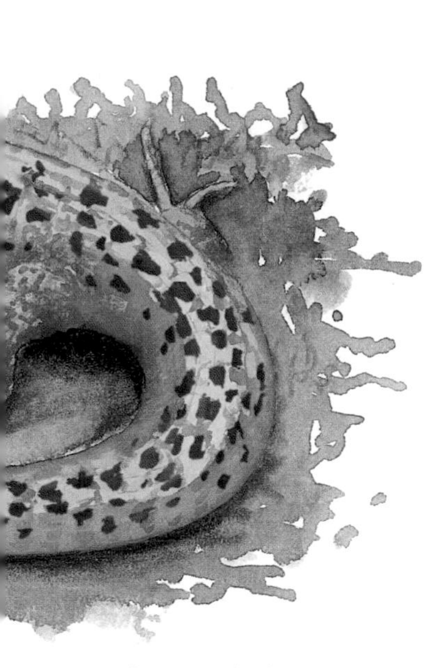

favourite basking site. If you scare one, return the next day more cautiously for a better look.

ID: brown upper body with small, dark blotches; white to greyish yellow belly; lateral scales are much smaller than those on back, similar to the scale pattern on an alligator; fold of skin on each side of body; tail makes up approximately half the length of body.

LENGTH: 15–25 cm (6–10").

DISTRIBUTION: from the Creston area of southern British Columbia to Vancouver Island. *Selected Sites:* Syringa Creek PP, southeastern Vancouver Island (both in BC).

HABITAT: savannah forests and grassland regions; under litter, bark or rocks; occasionally basks on rocks.

ACTIVITY PATTERNS: usually emerges from hibernation in April or May; remains active until September or October; more tolerant of cold than other lizards and may forage in colder temperatures.

REPRODUCTION: mates from April to June; female retains the eggs within her body; 3–7 live young are born in late summer.

FOOD: terrestrial invertebrates including slugs, spiders and worms.

SIMILAR SPECIES: *Western Skink* (p. 50): overall smooth and shiny appearance; broad, brown dorsal stripe; white stripe along both sides. *European Wall Lizard* (p. 106): confined to the Victoria region; collar of rough scales.

FRENCH NAME: Lézard-alligator boréal

DID YOU KNOW? The Northern Alligator Lizard has adapted to cold environments and can be found at an altitude over 3100 metres (10,000 feet) within its range. Another interesting adaptation is its ability to drop its tail to distract a predator and then later regrow the tail.

Five-lined Skink

Eumeces fasciatus

Among the fallen leaves of deciduous hardwood forests in a handfull of southern Ontario parks, Five-lined Skinks swim gracefully through the accumulated leaf litter, led by their finely tuned noses and propelled by their hyperdrive legs. Ontario's only lizard is often seen basking in the sun, but for most of its existence it prefers the sanctuary of leaf-littered ground.

• Woody debris, from rotting logs to boards, is crucial for Five-lined Skinks, as they are prone to water loss as adults and while in the egg. Studies in southern Ontario have shown that when logs and boards are removed, the population of skinks declines significantly. This impact has been felt most significantly in areas of high human disturbance along the northern shoreline of Lake Erie. Beach-loving sun-worshippers in particular not only remove or burn unsightly driftwood, but also trample vegetation. Skinks, unlike humans, prefer vegetated beaches scattered with logs—not featureless sand. Fortunately, efforts made to reclaim certain beaches to meet skink standards have resulted in a subsequent increase in numbers.

DID YOU KNOW? Although they are ground-lovers, Five-lined Skinks will sometimes be found hanging out or hunting in trees. Five-lined Skinks are not that common in Canada. Currently they are listed as Special Concern by COSEWIC, but they are likely to be up-listed to Threatened or Endangered in the southwestern Ontario portion of their range.

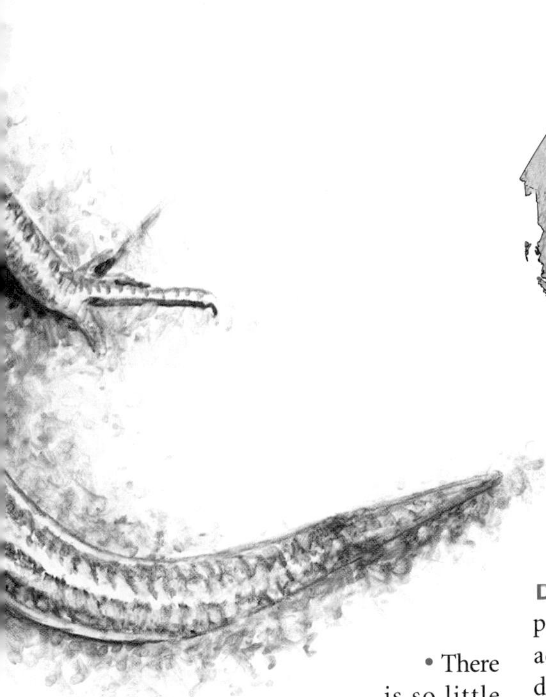

• There is so little offensive about Five-lined Skinks that their endearing qualities actually cause them additional harm. The pet trade has discovered these fine little lizards, and unfortunately, profiteers are still known to collect these animals even within Ontario's parks. No skinks should ever be removed from the wild because they are federally listed as vulnerable, and we need all the wild skinks we have. You will realize their value should you ever be so fortunate as to encounter a "Five-liner" in the Canadian wild.

ID: shiny, smooth and sleek, dark-coloured body; whitish belly; 5 cream-coloured stripes down back. *Breeding male:* bright orange chin. *Adult male and old female:* overall bronze colour; no stripes. *Female and immature:* blue or grey tail.

LENGTH: 13–21 cm (5–8").

DISTRIBUTION: a few small, isolated populations in southern Ontario, and across the southern edge of the Canadian Shield from Georgian Bay to the St. Lawrence River. *Selected Sites:* Murphy's Point PP, Lake Simcoe, Point Pelee NP, Pinery PP (all in ON).

HABITAT: open woodlands, sand dunes and rock outcrops that provide sufficient cover.

ACTIVITY PATTERNS: active from early April to mid-October; most active on warm, sunny days.

REPRODUCTION: breed in May and early June; females lay 2–15 eggs in a nest under cover or within a rotting log; females brood the eggs for up to 6 weeks; several females may nest at the same site; eggs hatch in late July.

FOOD: aquatic and terrestrial invertebrates.

SIMILAR SPECIES: Five-lined Skinks are the only lizards found in Ontario.

FRENCH NAME: Scinque pentaligne

Prairie Skink

Eumeces septentrionalis

Few people expect to see a lizard while walking through Manitoba's sand country, and the few who do might quickly dismiss it for a swift tiger beetle, grasshopper or sand-induced mirage. But for the determined lizard-lookers who know the Prairie Skink is indeed a Canadian reality, sightings of this sand swimmer can be cause for great satisfaction. • The Canadian population of the Prairie Skink is at the northern limit of its range. Only a few magical pockets of tall-grass sandy prairie worthy of exploration remain along the Assiniboine River, and it is here that the Prairie Skink can be found. These Manitoba populations are isolated from each other by surrounding croplands and are similarly separated from North Dakota populations. Owing to isolation, Canadian populations of the Prairie Skink are considered highly susceptible to potential disturbance and habitat loss, a particular concern as their open habitat is being quickly overgrown by aspen forests or converted to farmland. • The Prairie Skink

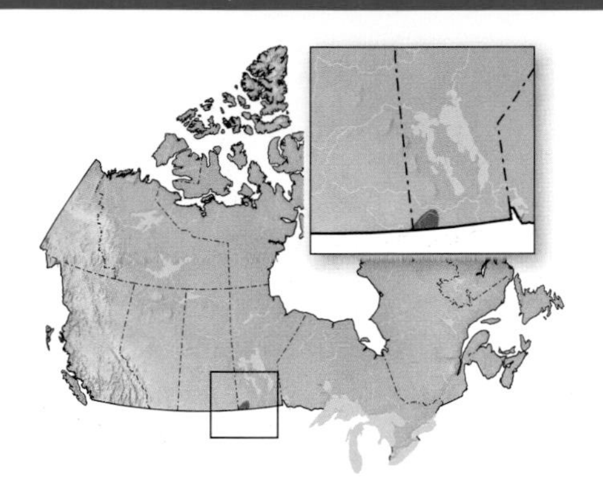

is an active hunter that prefers sun to clouds, but the odds of finding one under the warm Manitoba skies are not in your favour. It moves in jerky stops and starts when it chooses to leave its sanctuary beneath bark, branches and rocks, where it is most likely to be encountered. Salamander-seekers experienced in the ways of rock and log flipping are an asset to any Prairie Skink search party, but the lizard's getaway speed can still prevent prolonged encounters.

ID: brown body; 7 tan stripes on back and tail. *Breeding male:* orange on top of head. *Juvenile:* bright blue tail.

LENGTH: 13–22 cm (5–8 1/2").

DISTRIBUTION: only in sandy areas along the Assiniboine River. *Selected Sites:* Spruce Woods PP, Shilo Military Base (both in MB).

HABITAT: sandy areas with lots of woody debris, or on the Canadian Shield on rock outcrops; soft sand allows the skink to burrow down sufficiently into the ground to avoid extreme temperatures; woody debris and rock outcrops provide cover.

ACTIVITY PATTERNS: operates only during very warm weather; active from late April until early October; during hottest days of summer it will burrow into cooler sand.

REPRODUCTION: breeds in mid-spring; female lays 5–18 eggs; female broods eggs until they hatch in mid-summer.

FOOD: terrestrial invertebrates.

SIMILAR SPECIES: no other lizards are found in Manitoba.

FRENCH NAME: Scinque des prairies

DID YOU KNOW? Although skinks are not the most conspicuous of the world's lizards, there are more species in the skink family than in any other lizard family. The Prairie Skink is listed as Endangered by COSEWIC.

Western Skink

Eumeces skiltonianus

Away from the orchards and heavy development of the Okanagan Valley in southern British Columbia hides the reclusive Western Skink. It seeks more secretive terrain than other local reptiles that are more predictably found on scorched, sunny slopes. This skink is most commonly found by the logs, rocks and loose leaves between which it navigates life in the woodlands. The Western Skink emerges in March or April with the warm, spring sun and remains active until the onset of freezing weather, typically in October. It is best looked for on perfect summer days, but be prepared for a long search—the Western Skink is rarely seen. • A young skink, the kind most often encountered in searches, has an obvious, sky blue tail. This feature provides a distraction that predators or human searchers can focus on. Should the tail be grasped, it breaks off and wiggles chaotically, giving the lizard precious seconds of diversion in which to make an escape. The tail does regrow in time, and when a young skink reaches

SIMILAR SPECIES

Northern Alligator Lizard, p. 44

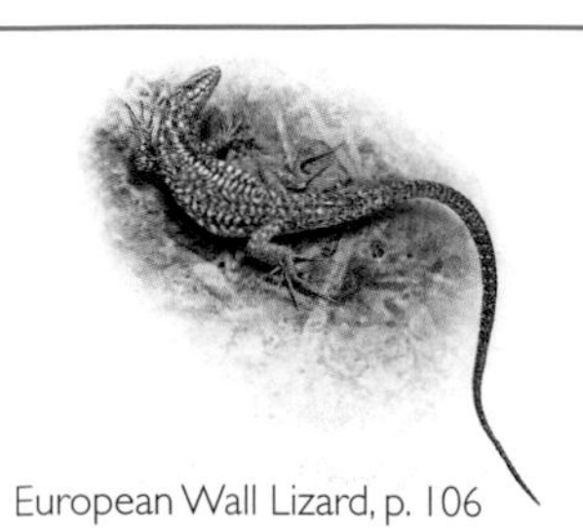

European Wall Lizard, p. 106

adulthood
the tail blends in
with the grey body.

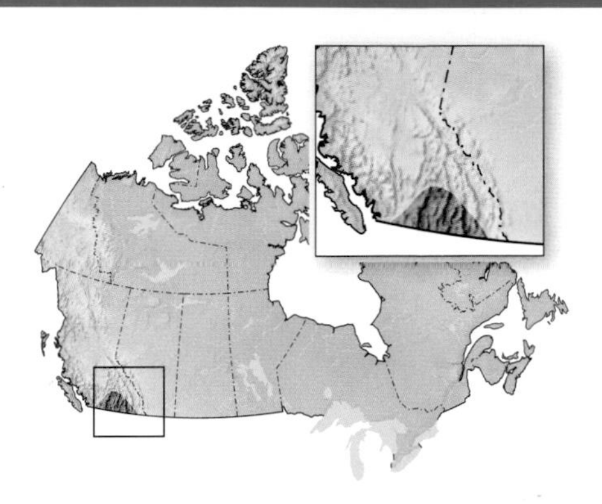

ID: slim body; tail longer than body; light brown stripe down back bordered by whitish stripe on each side; black sides; grey tail. *Breeding male:* orange on sides of head during breeding season. *Juvenile:* blue tail.

LENGTH: 15–18 cm (6–7").

DISTRIBUTION: south central British Columbia, particularly the southern Okanagan Valley. *Selected Sites:* Syringa Creek PP, Vaseux Lake PP, undeveloped talus slopes around Osoyoos (all in BC).

HABITAT: found in habitat that offers plenty of cover in the form of logs and rocks; can be found in various dry habitats, including drier forest types; tends to prefer forested areas more so than other lizards in British Columbia.

ACTIVITY PATTERNS: active during the day; will bury itself in sandy soil during the cold months.

REPRODUCTION: mates in spring; female lays 3–4 eggs under rocks, in crevices or burrows; female broods eggs until they hatch in July or August.

FOOD: terrestrial invertebrates.

SIMILAR SPECIES: *Northern Alligator Lizard* (p. 44): not as smooth looking; stripes on the underside; in juveniles the tail is the same colour as the body. *European Wall Lizard* (p. 106): currently confined to Vancouver Island.

FRENCH NAME: Scinque de l'ouest

DID YOU KNOW? In some areas, the Western Skink can be found at an elevation up to 2133 metres (7000 feet)! It is listed as Special Concern by COSEWIC.

Rubber Boa
Charina bottae

The Rubber Boa is a sluggish, inoffensive, small snake that looks like it is made of rubber or plasticine. This snake is a critter of darkness that rarely abandons its moist, cool daytime retreats to move about in the sun. The tiny, squinty eyes give further evidence to this fact, and so too does the lack of observations by people living within the heart of boa country. On rare occasions when the Rubber Boa leaves the safety of its underground burrows, it takes to the trees to hunt. Its arboreal feats are not as graceful as some other accomplished tree snakes; nevertheless, it proceeds with determination. • The differences between the boa and other Canadian snakes are not limited to physical appearance and activity patterns—they extend to capture of prey as well. Like the famed tropical boas popularized by jungle films, Canada's only native boa also constricts its prey. The Rubber Boa

SIMILAR SPECIES

Racer, p. 54

Sharp-tailed Snake, p. 56

seizes small animals and then methodically wraps itself tightly around the prey and asphyxiates it.

ID: "rubbery" body; uniform olive green, reddish brown or brown body; blunt tail; overall heavy body; small eyes with vertical pupils; anal spurs are found alongside the cloaca and are longer and curved on males. *Juvenile:* body may be pink.

LENGTH: 35–80 cm (14–32").

DISTRIBUTION: across much of southern British Columbia and north in the Interior. *Selected Sites:* the Kootenays and the Okanagan Valley (both in BC).

HABITAT: humid, mountainous habitats, usually under rocks or logs; may construct burrows.

ACTIVITY PATTERNS: in British Columbia, likely active from April to late October; during that period, primarily active at night, but a pregnant female will bask in the sun.

REPRODUCTION: breeds in May; female gives birth to 2–8 live young in late summer.

FOOD: small rodents and shrews; also salamanders, birds and lizards.

SIMILAR SPECIES: *Racer* (p. 54): belly is always yellow; thinner overall appearance; moves very quickly. *Sharp-tailed Snake* (p. 56): rust-coloured to grey body; black and white markings on belly; pointed tail.

FRENCH NAME: Boa caoutchouc

DIDYOU KNOW? When threatened, the Rubber Boa has been observed rolling itself up into a ball, tucking in its head and raising its tail as a head-like decoy. The Rubber Boa is listed as Special Concern by COSEWIC.

Racer

Coluber constrictor

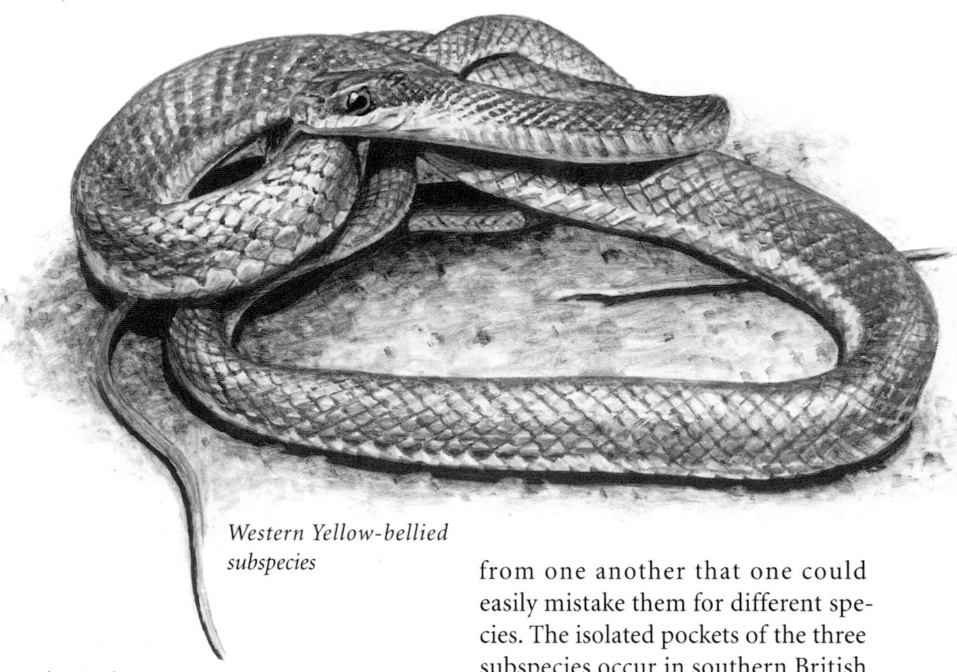

Western Yellow-bellied subspecies

The Racer is the sprinter of Canadian snakes. And not only is this agile snake fast, but it is also defensive. When caught, a Racer will bite repeatedly and nastily. It is therefore lucky for both the snake and people that the Racer's first line of defence is a speedy getaway. Most commonly, a Racer is seen or heard briefly as it retreats through the grass, seeming to disappear an instant later. • Canada has three different subspecies of the Racer, each so different from one another that one could easily mistake them for different species. The isolated pockets of the three subspecies occur in southern British Columbia, Saskatchewan and on Pelee Island in Ontario, and each subspecies is at its northernmost range. This is a snake of southern climates that struggles to hold on to the fringe of its range where habitat destruction and other harmful elements threaten eradication of its northern habitat. Human efforts to counteract society's development have helped the Blue Racer remain in Ontario, albeit only on a small island in Lake Erie. Similar efforts may be required for the other

SIMILAR SPECIES

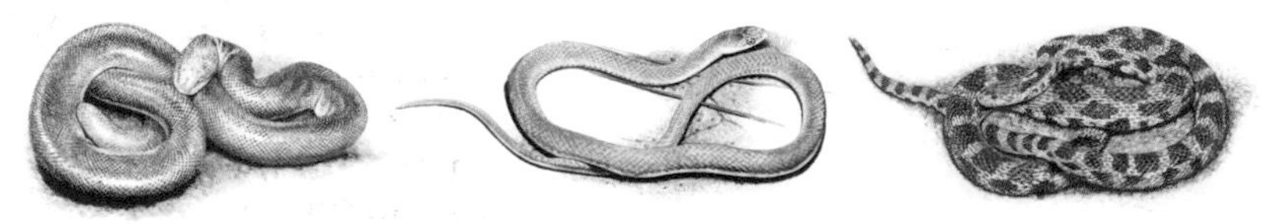

Rubber Boa, p. 52 Smooth Greensnake, p. 74 Eastern Foxsnake, p. 60

two populations of Canadian Racers, and although these are still fairly secure, continuing loss of habitat to agriculture, urban development and the oil and gas industry are threatening their future.

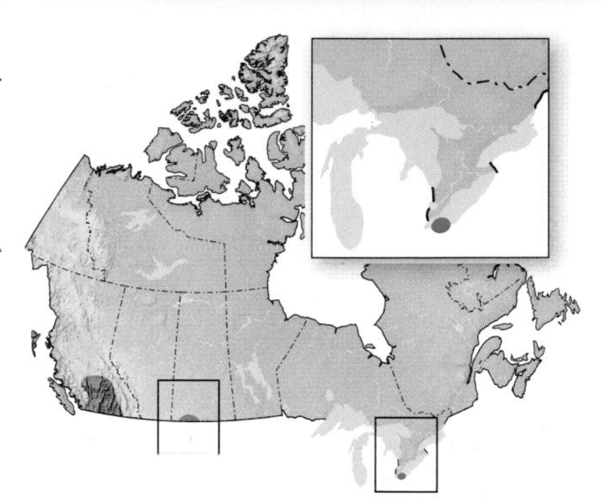

ID: very large eyes; carries its head and neck elevated, particularly when disturbed. In British Columbia, the *Western Yellow-bellied Racer*: green, beige or rufous upperparts; yellowish belly. In Saskatchewan, the *Eastern Yellow-bellied Racer:* light bluish green to grey or brown upperparts; light beige to yellow belly. On Pelee Island, the *Blue Racer*: light bluish green upperparts; white to light blue belly. *Juvenile:* distinctive mottled, black and white pattern that fades as it matures.

LENGTH: 60 cm–2 m (2–7').

DISTRIBUTION: *Western Yellow-bellied Racer*: drier parts of southern interior British Columbia. *Eastern Yellow-bellied Racer:* extreme southwestern Saskatchewan. *Blue Racer:* Pelee Island in Ontario. *Selected Sites:* Okanagan Valley grasslands (BC), Kalamalka Lake PP (BC), Grasslands NP (SK), Pelee Island (ON).

HABITAT: in British Columbia and Saskatchewan, grazed pastures, grasslands, savannah and rocky slopes; in southern Ontario, woodlands, abandoned fields and hedgerows.

ACTIVITY PATTERNS: hunts primarily during the day; emerges on warm days in April and enters hibernation in October.

REPRODUCTION: courtship and breeding occur shortly after spring emergence; female lays 3–9 eggs in a nest located in a rotting log, under rocks or in an animal burrow; eggs hatch by late summer; mature in 2–3 years.

FOOD: small rodents, birds and insects; pins its prey to the ground until it can ingest it.

SIMILAR SPECIES: British Columbia: *Rubber Boa* (p. 52): never has a yellow belly; "rubbery" skin. Saskatchewan: *Smooth Greensnake* (p. 74): brighter green throughout body. Pelee Island: *Eastern Foxsnake (juvenile)* (p.60): brown blotches; darker background colour.

FRENCH NAME: Couleuvre agile

DID YOU KNOW? On Pelee Island, the Blue Racer has successfully wintered in an artificially created hibernaculum and laid eggs in an artificial nest site. COSEWIC has listed all three subspecies as being at risk: the Western Yellow-bellied Racer is Special Concern; the Eastern Yellow-bellied Racer is Threatened; and the Blue Racer is Endangered.

Sharp-tailed Snake

Contia tenuis

Seven sites on Vancouver Island and the neighbouring Gulf Islands have yielded this unusual snake, hinting at its restricted nature. It seems that the Sharp-tailed Snake has a very patchy distribution and that in each of these sites, it is only found within a two-kilometre- (1.2-mile-) diameter core area. Some habitat fragmentation is partly responsible for this restriction, but also, this species occurs at the northern reaches of its physiological limits, and the cool temperatures restrict its distribution. • This animal is a strange snake indeed. For one thing, it seems to specialize in eating slugs. However, one of the most

SIMILAR SPECIES

Terrestial Gartersnake, p. 86

Northwestern Gartersnake, p. 88

Common Gartersnake (Red-spotted subsp.), p. 94

commonly seen and conspicuous slugs, the Banana Slug, is too large to be made a meal of by this delicate snake. Other smaller types of slugs are known to occur, of course, but most are introduced. The native coastal slug population has been largely replaced by immigrant relatives. It is not known if this big change in prey base has affected the populations of the Sharp-tailed Snake in Canada. • There have been very few people who have had the pleasure of encountering this animal in the wilds of British Columbia. For this reason, very little is known about its needs and activities at the northern edge of its range. From studies elsewhere we have learned that the Sharp-tailed Snake is frequently found under cover in oak, arbutus or fir forests or has occasionally dug deep into a rotten log.

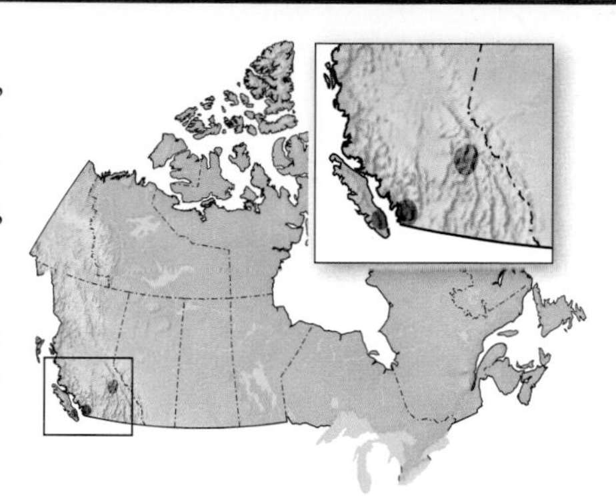

ID: very small; rust-coloured to grey body; black and white barring on belly; tail is sharp, ending in a pointy spine.

LENGTH: 20–45 cm (8–18").

DISTRIBUTION: most Sharp-tailed Snakes have been found on Vancouver Island and the Gulf Islands; there is one possible record from the British Columbia interior as well. *Selected Sites:* southeastern Vancouver Island and the neighbouring Gulf Islands (both in BC).

HABITAT: relatively open Douglas-fir–Arbutus forest edges on south-facing, rocky slopes.

ACTIVITY PATTERNS: may be nocturnal, as it feeds on slugs and is active at relatively low temperatures; becomes active in early spring and again in late October or early November; aestivation likely occurs during the dry summer months; hibernates underground in winter.

REPRODUCTION: very little is known; female lays 2–5 eggs between June and early July; eggs hatch in approximately 4–6 weeks; young are around 7.5 cm (3") long.

FOOD: predominantly slugs; other invertebrates and the occasional salamander.

SIMILAR SPECIES: *Terrestial* (p. 86), *Northwestern* (p. 88) and *Common* (p. 94) *gartersnakes*: usually boldly patterned, but juvenile gartersnakes may look similar without close inspection.

FRENCH NAME: Couleuvre à queue fine

DID YOU KNOW? The Sharp-tailed Snake has extremely long teeth—not fangs like in rattlesnakes, but long teeth in the rear to puncture and grasp slimy slugs. The function of the sharp tail is still not known. The Sharp-tailed Snake is listed as Endangered by COSEWIC.

Ring-necked Snake

Diadophis punctatus

You have bigger problems than mere ophidiophobia if you are finding yourself afraid of the dainty Ring-necked Snake. Although the fear of snakes probably has some deep-rooted evolutionary significance, these pencil-sized wigglers are about as fearsome as the plastic toy store models collected by youngsters; in fact, you are in more physical danger from the poison ivy and twig scratches that may afflict you while searching for this snake than from whatever it can muster. • For the naturalist, most encounters occur when out looking for salamanders. The Red-backed Salamander is the most frequent under-rock inhabitant (aside from ants, sowbugs and beetles), and it is encountered many more times

SIMILAR SPECIES

Red-bellied Snake, p. 82

Queen Snake, p. 78

than this little snake. The Red-backed Salamander is also a favourite snack of the Ring-necked Snake. • It is unfortunate that most Canadians will never see a Ring-necked Snake—indeed many Canadians likely live within an hour or two from these snakes. Experiencing one in the wild not only confirms that a snake can actually be cute, but also that most snakes need not be feared.

ID: small and slender; defined yellow, cream or orange ring around neck; grey to black body; yellow, orange or red belly.

LENGTH: 25–60 cm (10–24").

DISTRIBUTION: from southern Ontario to the Maritimes. *Selected Sites:* Georgian Bay region, Murphy's Point PP, Charleston Lake PP, Algonquin PP (all in ON).

HABITAT: wooded areas and nearby meadows; surface bedrock and organic debris.

ACTIVITY PATTERNS: primarily active at night between the months of April and October.

REPRODUCTION: breeds in spring or fall; in early summer, female lays 2–6 eggs in rotting logs or other debris; several females may lay their eggs in one nest; eggs hatch in about 2 months; young reach sexual maturity in 2–3 years.

FOOD: Red-backed Salamanders and other small vertebrates; invertebrates such as beetles, worms and arthropods living in damp leaf litter.

SIMILAR SPECIES: *Red-bellied Snake* (p. 82): at most may have a few spots around neck. *Queen Snake* (p.78): lacks clear neck ring; 4 dark lines along belly.

FRENCH NAME: Couleuvre à collier

DID YOU KNOW? When disturbed, the Ring-necked Snake will usually expose its colourful underside and occasionally discharge a smelly musk to deter the perceived threat.

Eastern Foxsnake

Elaphe gloydi

More than 70 percent of all Eastern Foxsnakes are found in Ontario. What is also found in this same region is a particularly dense zone of humans, industrial and agricultural development and pollution. These persistent threats have greatly contributed to a significant decline in snake numbers and a corresponding ascension onto lists of endangered wildlife. Fortunately, the threatened Eastern Foxsnake has some of its habitat protected: it is found in two national parks and more than 15 provincially protected areas. • This snake is large, second in size only to the closely related Eastern Ratsnake. Like its relative, it has strong tree-climbing instincts and can also be found undulating through marshes or lakes and even swimming, especially among islands in Georgian Bay. However, the Eastern Foxsnake seems most comfortable on the ground, where its patterned colours blend nicely with fallen leaves or dried grasses. This camouflage hides it from most passing eyes, but should intruders come too close, it will vibrate its tail in surface litter, making a noise similar to

SIMILAR SPECIES

Northern Watersnake, p. 72 Milksnake, p. 70 Eastern Hog-nosed Snake, p. 66

that of a perturbed rattler. This defensive strategy most often has the desired outcome of scaring off the threat, but this snake is certainly not one to be feared: it is not venomous and, as it is a constrictor, doesn't even have a taste for biting—it is simply a gentle giant among Canada's snakes.

ID: yellow to light brown body; large, dark brown spots down centre of back; 2 alternating rows of smaller, dark brown spots along sides; white to cream belly covered with dark specks. *Male:* proportionally longer tail than female.

LENGTH: 90 cm–1.7 m (3–6').

DISTRIBUTION: southern Ontario; confined to the shoreline and islands of Georgian Bay, the shoreline and islands of western Lake Erie and Essex and Kent counties. *Selected Sites:* Long Point PP, Rondeau PP, Point Pelee NP, Killbear PP, Georgian Bay Islands NP (all in ON).

HABITAT: prefers wet areas; found in mainly unforested areas such as fields, farmland, lake shorelines, rolling prairies, wooded stream valleys and exposed rock.

ACTIVITY PATTERNS: primarily active during the day; can also be found on warm nights in summer and fall; leaves a winter den in April and returns to hibernate in late October or early November.

REPRODUCTION: mates from May to June; female deposits 15–20 eggs under logs or boards or in organic piles; females may share nest sites; young hatch in late August to October and are 25–30 cm (10–12") long.

FOOD: voles, mice, bird eggs, fledgling birds and young rabbits.

SIMILAR SPECIES: *Northern Watersnake* (p. 72): does not have bold blotches. *Milksnake* (p. 70): dark halos around its dark markings. *Eastern Hog-nosed Snake* (p. 66): upturned snout; much smaller. *Massasauga* (p. 100): distinct rattle; darker base colour; lighter blotches.

FRENCH NAME: Couleuvre fauve de l'est

SIMILAR SPECIES

Massasauga, p. 100

DID YOU KNOW? On Pelee Island, an Eastern Foxsnake nest was found in the same rotting log as the nests of a Blue Racer and a Snapping Turtle. Eastern Foxsnakes are also known to share their hibernaculum with other kinds of snakes. The Eastern Foxsnake is listed as Threatened by COSEWIC.

Eastern Ratsnake

Elaphe obsoleta

This poor snake's name makes it seem like a vile critter, lurking through city sewers and town dumps for a tasty rat. This perception is wrong. Most often when Eastern Ratsnakes are encountered, it is in the wild, fragrant areas of southern Ontario, among bird songs and green leaves. • Admittedly, Eastern Ratsnakes are intimidating creatures when first seen. Many Eastern Ratsnakes grow longer than people are tall, a trait that seems not to inconvenience them in trees—their most frequent hangout. Up in branches, Eastern Ratsnakes actively hunt and constrict their prey; they also find shelter and basking sites. It is now thought that prime basking sites may influence where these snakes communally den in the winter. Of course, the snakes do not bask during the season of snow, but when spring rolls in they are committed to worshipping the sun. • Despite their tree-dwelling nature, Eastern Ratsnakes descend to the ground and cover quite a bit of territory during summer. Genetic tests reveal that despite the limited distribution in Ontario, the gene pool is diverse principally because snakes move from population to population, spreading their genetic traits far and wide.

SIMILAR SPECIES

Northern Watersnake, p. 72

ID: mostly black with some white that shows between the scales; whitish chin

and throat; white belly with black markings. *Hatchlings and Juveniles:* strongly marked with a dorsal row of brown patches and smaller side spots on a pale grey background.

LENGTH: 1–2.6 m (3–9').

DISTRIBUTION: 5 isolated spots on the north shore of Lake Erie and a larger population north of Kingston on the eastern end of Lake Ontario. *Selected Sites:* Murphy's Point PP, Charleston Lake PP, St. Lawrence Islands NP, Long Point PP, Frontenac Axis, Queen's University Biological Station (all in ON).

HABITAT: mature, upland hardwood forests and swamps; may also be found in abandoned fields and meadows; during early summer are commonly found along forest edges; hibernate communally on rocky, south-facing slopes in natural fissures or tree root systems.

ACTIVITY PATTERNS: active during the day between mid-April and early October.

REPRODUCTION: breed shortly after emerging in spring; females will lay 12–16 eggs once every 2–3 years in a rock pile, decaying stump, tilled soil or organic pile; young snakes hatch by fall.

FOOD: small mammals, birds and nestlings.

SIMILAR SPECIES: *Northern Watersnake* (p. 72): generally patterned, but older snakes are darker; belly is whitish with dark, moon-shaped markings; lacks white throat.

FRENCH NAME: Couleuvre obscure

DID YOU KNOW? Male and female Eastern Ratsnakes along the Frontenac Axis are thought to reach sexual maturity when they are 10–12 years old. They may live up to 25 years in the wild. Eastern Ratsnakes are listed as Threatened by COSEWIC.

Western Hog-nosed Snake

Heterodon nasicus

Never far away from sand or mystery, the Western Hog-nosed Snake is known to few and seen by even fewer. Looking for a Western Hog-nosed Snake on the grasslands of Alberta or Saskatchewan more often results in a pleasurable walk than a snake sighting. The trick to finding one may be in searching out this snake's favourite food—toads. The Western Hog-nosed Snake has specialized teeth in the back of its mouth to pop toads that inflate themselves as their defence. Its shovel-like nose

SIMILAR SPECIES

Gophersnake, p. 76

Prairie Rattlesnake, p. 98

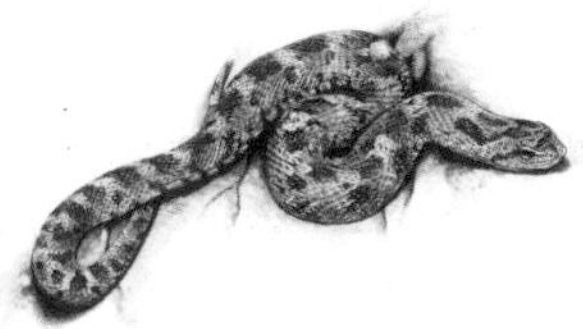

Eastern Hog-nosed Snake, p. 66

allows it to uproot toads from their underground burrows after using its keen sense of smell to detect them. So hanging out at your local prairie pond may be your best bet to see a Western Hog-nosed Snake. • This snake is the western counterpart of the Eastern Hog-nosed Snake. Both snakes share the strange defence of rolling over and playing dead, even hanging out their tongues for special effect; however, the Western Hog-nosed Snake does this less than the other members of the Hog-nosed Snake family.

ID: thick-bodied; pointy, upturned snout; light brown or grey base colouring; dark blotches down back; 2 or 3 rows of smaller spots along sides; belly is patterned with dark blotches.

LENGTH: 40–90 cm (16–36").

DISTRIBUTION: the southern parts of the prairie provinces. *Selected Sites:* Suffield NWA (AB), Grasslands NP (SK), Spruce Woods PP (MB).

HABITAT: prefers sandy areas on the arid grasslands; floodplains; scrubby areas.

ACTIVITY PATTERNS: active during the early morning and evening hours; burrows into the ground by day; likely hibernates from October to April.

REPRODUCTION: probably mates through spring; clutches of 4–23 eggs are laid in sandy soil during summer; young hatch after 2 months; maturity is reached in 2 years; may live up to 8 years.

FOOD: primarily toads; also birds, mammals and invertebrates depending on the size of the snake; venom is toxic to prey but harmless to humans.

SIMILAR SPECIES: *Gophersnake* (p. 76) and *Prairie Rattlesnake* (p. 98): snout not upturned. *Eastern Hog-nosed Snake* (p. 66): does not occur on the prairies.

FRENCH NAME: Couleuvre à nez retroussé

DID YOU KNOW? When this snake strikes at an intruder, it is almost always done with a closed mouth. It may also inflate its neck to look bigger. Rolling over and playing dead makes it an easy target for a shovel or other weapon. Clearly these evolved defences don't work well against human threats.

Eastern Hog-nosed Snake

Heterodon platirhinos

Fear does strange things. There are many cases of snake-human encounters wherein fear is so overpowering that during the meeting one party rolls onto its back, sticks out its tongue and looks to all the world like it has died. This act is the common response of the Eastern Hog-nosed Snake to our presence. If you think you're queasy of snakes, just think of how disturbing you must be to an Eastern Hog-nosed Snake! Its death-feigning trait is legendary among snake-lovers across the continent. However, when initially disturbed, the Eastern Hog-nosed Snake can put on a pretty good show of self-defence involving a neck-spreading cobra impersonation and repeated closed-mouth strikes. It is only if this feisty strategy fails that the snake opts for the "roll over and play dead" approach. • This snake's unusual snout, tipped by an upturned scale, serves as a built-in shovel. Nosing into soft sand, the Eastern Hog-nosed Snake finds not only sanctuary from the heat, but also buried treasures of food. Toads that seek refuge in sand are rooted out by this excavator. Once grasped, the prey is punctured by rear fangs and slowly immobilized by toxic saliva. • Unfortunately, its ability to stalk the sands of southern Ontario is being continually

SIMILAR SPECIES

Northern Watersnake, p. 72 Milksnake, p. 70 Eastern Foxsnake, p. 60

threatened by shoreline development because beaches are also very popular with people. Thus, loss of habitat, increased roadkill and persecution by people have led to the disappearance of this fascinating snake from much of its former range.

ID: thick body; wide neck; distinctive, upturned snout; colour can range from yellow to light brown (rarely to black), patterned with large, dark blotches; underside is pale; tail is lighter than belly.

LENGTH: 50 cm–1.2 m (20"–4').

DISTRIBUTION: in Canada, restricted to southern Ontario. *Selected Sites:* Lake Simcoe region, London to Owen Sound, Bachus Woods, Georgian Bay, Wasaga Beach PP (all in ON).

HABITAT: prefers sandy regions such as those along the shorelines of Lake Erie, Georgian Bay and Lake Ontario; may also be found in pine or oak woodlands where loose soil and a source of water exist.

ACTIVITY PATTERNS: active during the day and on warm nights; overwinters singly or in small groups from October to mid-spring.

REPRODUCTION: mating follows emergence in spring; in late June or early July female lays about 20 eggs in rotting logs, rocks or sandy soil; eggs generally hatch in 8 weeks; young are approximately 20 cm (8") long; young will become sexually mature after 2 years.

FOOD: toads and, occasionally, other amphibians.

SIMILAR SPECIES: *Northern Watersnake* (p. 72), *Milksnake* (p. 70) and *Eastern Foxsnake* (p. 60): all have blotches, but none have an upturned snout. *Western Hog-nosed Snake* (p. 64): not found in Ontario.

FRENCH NAME: Couleuvre à nez plat

SIMILAR SPECIES

Western Hog-nosed Snake, p. 64

DID YOU KNOW? Even if an Eastern Hog-nosed Snake is flipped over while playing dead, it will roll over onto its back again! The Eastern Hog-nosed Snake is listed as Threatened by COSEWIC.

Nightsnake
Hypsiglena torquata

Darkness is the sanctuary of the Nightsnake. Whether under the shelter of a fallen tree or board, it is under these conditions that curious snake lovers must seek to meet a Nightsnake. Even armed with this information, most eager naturalists will fail in their search, primarily because there aren't that many Nightsnakes to be found in Canada. In fact, since their first Canadian discovery in 1980, only about one a year has been encountered in the wild. The Nightsnake of the Okanagan Valley ties Canada to the Great Basin deserts. This shrub desert environment is about as near to Tucson, Arizona, that you can find in Canada, and that's apparently just good enough for this snake. It has not been found more than 50 kilometres (31 miles) from the American border. • The Nightsnake seems to prefer amphibians and reptiles over a diet of small mammals and birds. A look into the snake's mouth might just explain this preference. The snake's rear teeth are bigger and slightly grooved in comparison to its other teeth. These rear fangs may allow the snake to hold on tight to slippery prey such as frogs and salamanders, allowing the mild venom in the snake's spittle to slowly snuff life from the prey. • As a result of the limited experiences with the Canadian Nightsnake, very little information is known about its natural history.

ID: beige to grey body; numerous darker blotches on back and sides; vertical

SIMILAR SPECIES

Western Rattlesnake, p. 96

Terrestial Gartersnake, p. 86

Gophersnake, p. 76

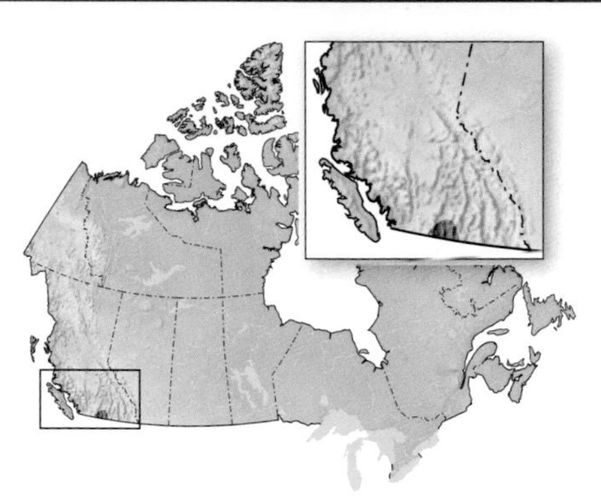

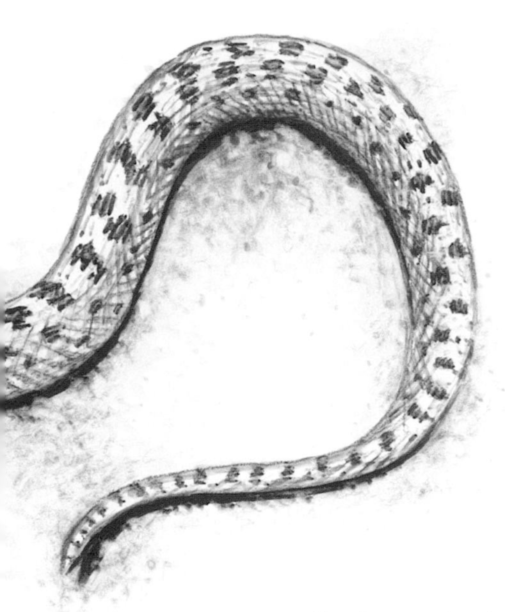

pupils; large, dark blotches on either side of neck; white belly.

LENGTH: 30–60 cm (12–24").

DISTRIBUTION: Okanagan Valley south of Penticton, British Columbia; area surrounding Keremeos, British Columbia. *Selected Sites:* southern Okanagan Valley, Similkameen Valley (both in BC).

HABITAT: hot, arid regions; found under rocks near talus slopes and under other debris such as boards or fallen trees.

ACTIVITY PATTERNS: active at night between May and October.

REPRODUCTION: little information is known; likely breeds in spring; female lays 4–6 eggs; eggs hatch in about 2 months.

FOOD: amphibians and lizards, sometimes small snakes; in Canada they likely consume large numbers of Western Skinks.

SIMILAR SPECIES: *Western Rattlesnake* (p. 96): rattle on end of tail; triangular head. *Terrestrial Gartersnake* (p. 86): tan or yellow stripe along back. *Gophersnake* (p. 76): larger; round pupils. *Racer* (p. 54): no blotches; lacks vertical pupils.

FRENCH NAME: Couleuvre nocturne

SIMILAR SPECIES

Racer, p. 54

DID YOU KNOW? A Nightsnake near Osoyoos, B.C., regurgitated a neonate Western Rattlesnake upon capture. It was estimated that the prey weighed in at close to half that of the predator. The Nightsnake is listed as Endangered by COSEWIC.

Milksnake

Lampropeltis triangulum

Fall in Ontario's Cottage Country involves a half-hearted preparation for winter during which sheds and summer debris are cleaned out—resulting in the occasional discovery of a colourful Milksnake. The Milksnake is one of the more common snakes to be found around human habitation. This constrictor takes advantage of our buildings and debris as cover and feeds upon small mammals that we unintentionally attract with our food stores. This association hints to the snake's name; it was once thought that the Milksnake entered barns to milk cows. In truth, the Milksnake has always had a fondness for barns, but with a taste for mice—not milk. This misconception shows how uninformed one can be about snakes. The Milksnake has small, needle-sharp teeth, and one can readily imagine the snake's fate if

SIMILAR SPECIES

Northern Watersnake, p. 72 Eastern Hog-nosed Snake, p. 66 Eastern Foxsnake, p. 60

it tried to attach itself to a cow's udder. • The Milksnake is a fine looking and completely harmless animal. It is closely related to the Corn Snakes and Kingsnakes that are commonly sold in pet stores. Although they are not quite as passive as the captive models, a wild Milksnake's fair temper does hint to this relation.

ID: grey to tan body; black-rimmed, brown to rusty blotches on back; Y- or V-shape on back of neck.

LENGTH: 60 cm–1.4 m (2–5').

DISTRIBUTION: southern Quebec and southern Ontario. *Selected Sites:* Cottage Country, Charleston Lake PP, Murphy's Point PP (all in ON).

HABITAT: found around forest clearings and edges and agricultural fields.

ACTIVITY PATTERNS: encountered during the day but tends to be more active in the evening; hibernates from October to mid-spring.

REPRODUCTION: mates soon after emerging in spring; female then lays 3–24 eggs in a nest in a small, abandoned mammal burrow or rotting log; eggs hatch in 7–10 weeks; individuals reach sexual maturity in 3–4 years.

FOOD: small rodents such as mice and voles; also birds, amphibians, reptiles and even large insects.

SIMILAR SPECIES: *Northern Watersnake* (p. 72): darker; not brightly patterned; lacks "V" or "Y" mark on top of head. *Eastern Hog-nosed Snake* (p. 66) and *Eastern Foxsnake* (p. 60): lack black outline around blotches.

FRENCH NAME: Couleuvre tachetée

DID YOU KNOW? Although it doesn't look like a rattlesnake, a disturbed Milksnake will shake the end of its tail in leaf litter to emulate the sound of its more dangerous relative. The Milksnake is listed as Special Concern by COSEWIC.

Northern Watersnake

Nerodia sipedon

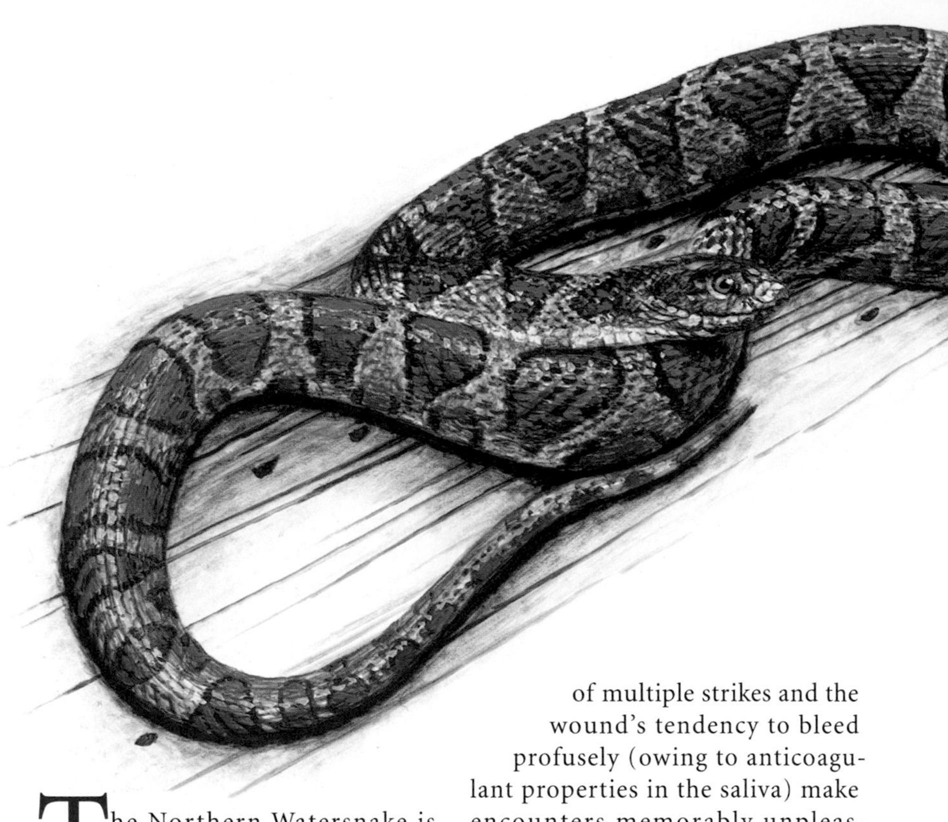

The Northern Watersnake is born feisty and just gets worse with age. This is of course an anthropocentric perception, but for just about any herpetologist that has ever attempted to catch one, tiny scars serve as a testament of the snake's temperament. Bites from this snake are not overly painful, but its habit of multiple strikes and the wound's tendency to bleed profusely (owing to anticoagulant properties in the saliva) make encounters memorably unpleasant. In truth, the perceived nastiness makes the Northern Watersnake even more interesting. • The most commonly seen large snake in Canada, it is a superb hunter of wetland margins, gracefully undulating through waters and even diving down to depths of three metres (10 feet) in pursuit of

SIMILAR SPECIES

Eastern Foxsnake, p. 60

Milksnake, p. 70

Eastern Hog-nosed Snake, p. 66

fishes. When not actively hunting, this mud-coloured snake can be found basking on waterside logs or rocks, metabolizing its meal. Meanness is reserved for threats and meals only, as these snakes bask and hibernate peacefully in groups. An unfortunate by-product of this conspicuous basking is unwarranted persecution or ignorant brutality by those without an appreciation for snakes. It is too frequently killed by those who try to pick up the grumpy snake or who mistake its identity—the Northern Watersnake is often thought to be a Water Moccasin, a venomous species that does not even occur in Canada. • Although the Northern Watersnake is one of the most commonly seen snakes around wet areas of Ontario and Quebec, it can be found far from water, likely in an effort to disperse to other wet areas.

ID: variable colouring; body becomes darker with age, eventually black; dark brown blotches that are wider than the lighter bands; dark bands may alternate colour on sides of some snakes; white belly or yellow belly with dark crescents. *Lake Erie Watersnake*: plain grey-olive above; not banded or blotched.

LENGTH: 60 cm–1.4 m (2–5').

DISTRIBUTION: south of Lake Superior in Ontario and southern Quebec. *Selected Sites:* Pelee Island (ON).

HABITAT: edges of freshwater lakes, marshes, swamps and slow-moving streams; strongly tied to water.

ACTIVITY PATTERNS: active through the day, but may rest and bask during the midday heat, sometimes in groups; hibernates for much of the year to avoid the cold.

REPRODUCTION: breeds in aggregations shortly after emerging from hibernation; female breeds with multiple partners—consequently not all of the young have the same father; female gives birth to 15–50 live young in August or September; sexual maturity is reached in 3–4 years.

FOOD: fishes and amphibians.

SIMILAR SPECIES: *Eastern Foxsnake* (p. 60): yellowish base colour. *Milksnake* (p. 70): Y- or V-shaped, light patch on back of head. *Eastern Hog-nosed Snake* (p. 66): upturned snout.

FRENCH NAME: Couleuvre d'eau

DID YOU KNOW? The Lake Erie Watersnake, a subspecies found only on islands in the western end of Lake Erie and part of Ohio, is endangered—the main threat is persecution by humans. Even though it is rare and endangered, this variant is readily seen along the rocky shores of Pelee Island.

Smooth Greensnake

Opheodrys vernalis

The Smooth Greensnake is an ideal literal example of "a snake in the grass." This small, green wiggler is most often found among grasses as it slithers swiftly through the tangle of blades. It isn't easy to see and is even harder to follow when it makes for a retreat. The pencil thin form blends so well into its surroundings that it seems to be made of the grass itself.

DID YOU KNOW? Shortly after the Smooth Greensnake dies, it loses its vivid green colour and turns blue. This may cause it to be mistaken for a Blue Racer. However, a Blue Racer this small would be a highly blotched juvenile.

When it comes to hiding, this snake has a deep bag of tricks, and the one used when hiding in low shrubs is particularly effective. While climbing among low branches, the Smooth Greensnake gently sways the front half of its body, mirroring the movements of leaves in the wind. The ruse is very effective at concealing it when it finds itself in a shrub, but the Smooth Greensnake prefers to be in the grass. • Whenever the snake leaves the grass, trouble looms near. In particular, the Smooth Greensnake is attracted to the hotplate surface of roads in the evening. Many are unknowingly killed by drivers, but nevertheless the Smooth Greensnake

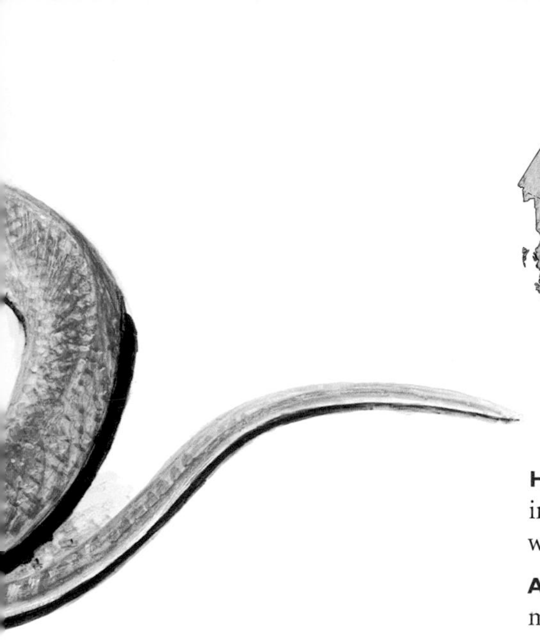

probably remains common, although not conspicuously so, across much of Canada.

ID: uniformly bright green body; white or yellowish belly; smooth scales. *Juvenile:* blue-grey to dark green body.

LENGTH: 30–60 cm (12–24").

DISTRIBUTION: from southeastern Saskatchewan to south-central Manitoba; from southern Ontario east of Lake Superior through southern Quebec to the Maritime provinces. *Selected Sites:* Cottage Country, Charleston Lake PP, Georgian Bay Islands NP (all in ON), Kejimkujik NP (NS), Gatineau Park (QC).

HABITAT: moist, grassy areas including wet meadows, marshes and open woodlands.

ACTIVITY PATTERNS: active primarily during the day between mid-spring and mid-fall.

REPRODUCTION: mates in spring or summer; female may "incubate" eggs within her body by basking in the sun, quickening embryonic development; female lays 3–11 eggs; nest may be found in a rotting log or animal burrow or under boards or other debris; eggs hatch in 1–3 weeks; young become sexually mature in 2–3 years.

FOOD: spiders, insects and other invertebrates.

SIMILAR SPECIES: no other native snake is bright green.

FRENCH NAME: Couleuvre verte

Gophersnake

Pituophis catenifer

All feelings of ophidiophobia (fear of snakes) aroused by Gophersnakes fade quickly once you get to know them. These snakes are large, but they are gentle giants. Initially, Gophersnakes (particularly young individuals) may show some aggression. Deep guttural hissing, tail shaking and the occasional half-hearted strike serve to dismiss unwelcome encounters. Should a meeting get beyond these first impressions, they soon calm down and become most pleasant. However, if continually harassed, these constrictors can give a painful but nonpoisonous bite. • Marauders of mice and other rodents, Gophersnakes prey heavily on animals considered pests by people that make their living off the land. Gophersnakes are extremely powerful and whack their prey with a full-on strike, then follow up with a strangling coil if needed. • These

SIMILAR SPECIES

Western Rattlesnake, p. 96

Prairie Rattlesnake, p. 98

Western Hog-nosed Snake, p. 64

snakes are well known for their ability to climb trees. In the cottonwood forests of prairie rivers, Gophersnakes cautiously go vertical in the hopes of finding bird nests. When they encounter one, the eggs or young are meticulously eaten. It is a shame that we do not appreciate the snakes' purposeful ways; Gophersnakes are too often killed by people who don't realize the range of values the snakes possess.

ID: large and heavy-bodied; boldly patterned with alternating dark brown or reddish, rectangular blotches on a cream or tan body; banding tends to be heavier toward tail; belly is cream coloured with a few dark marks; pupils are round; head is rounded.

LENGTH: 90 cm–2 m (3–7'); some individuals may be even longer!

DISTRIBUTION: southern Saskatchewan, Alberta and interior British Columbia. *Selected Sites:* Grasslands NP (SK), Dinosaur PP (AB), Writing-on-Stone PP (AB), Okanagan Valley (BC).

HABITAT: short-grass prairie; areas of sandy soil; scrubland, rock piles, native pastures and cottonwood forests.

SIMILAR SPECIES

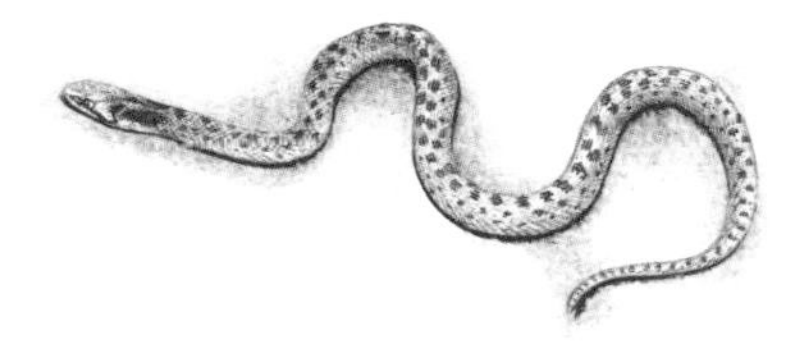

Nightsnake, p. 68

ACTIVITY PATTERNS: active during the day, although not in extremely hot weather; emerge in April and retreat to hibernation sites in October.

REPRODUCTION: mate once they emerge from hibernation; females lay 2–8 eggs once or twice during summer; eggs are often laid in a small, unused mammal burrow; hatchlings are 20–40 cm (8–16") long and appear in late summer or early fall.

FOOD: small rodents such as voles and mice; also birds, amphibians and invertebrates.

SIMILAR SPECIES: *Western Rattlesnake* (p. 96) and *Prairie Rattlesnake* (p. 98): triangular head; vertical pupil; rattle on end of tail. *Western Hog-nosed Snake* (p. 64): upturned nose; smaller on average. *Nightsnake* (p. 68): thinner; vertical pupils.

FRENCH NAME: Couleuvre à nez mince

DID YOU KNOW? Gophersnakes do not have a rattle, but they do vibrate their tail when agitated. When the tail is rattled in dry grass, it sounds similar the buzz of a rattlesnake.

Queen Snake

Regina septemvittata

The Queen Snake is a lover of water whose only royal connection is its refined palate. Like Her Majesty's "namesakes" worldwide, the Queen Snake has a particular fondness for delicacies, which in this case means crustaceans. And not just any shellfish: this snake feeds almost exclusively on soft-shelled crayfish that have just molted from their old shells. • Unfortunately, animals that snub their noses at most food and rely heavily on just one prey item give up the stable benefits of a generalist predator. This problem currently afflicts the Queen Snake, as its fortunes mirror those of the crayfish. Pollutants, shoreline developments around the Great Lakes and introduced crayfish directly affect crayfish habitat and have caused significant declines in population sizes of native crayfish throughout the area. Because of its habitat and dietary preferences, the Queen Snake is susceptible to pollutants. The snake is exposed to mercury toxicity via contaminated crayfish, and other water pollutants infiltrate its body through its highly permeable skin. • Effects from development are difficult for a naturalist in southern Ontario to ignore, but on occasion these depressing thoughts fade away when one goes searching for a Queen Snake in clean streams and marshlands. A typically royal desire

SIMILAR SPECIES

Northern Watersnake, p. 72

Butler's Gartersnake, p. 84

Common Gartersnake (Eastern subsp.), p. 94

for pristine surroundings on the part of the Queen Snake encourages naturalists to visit some of southern Ontario's last green jewels when searching for this snake.

ID: solid, dorsal, brownish colouring can range from tan to near black; yellow stripe low along side; 4 dark lines run along length of belly.

LENGTH: 40–90 cm (16–36").

DISTRIBUTION: southern Ontario in a few widely separated locations. *Selected Sites:* River Canard, Upper Thames River, Maitland River (all in ON).

HABITAT: requires slow-flowing or still, permanent waters that are at least 18° C (65° F) for most of the active season; flat rocks along the shoreline are also important features that provide cover; may also be found around marshes with a high water table.

SIMILAR SPECIES

Red-bellied Snake, p. 82

ACTIVITY PATTERNS: active primarily by day, but sometimes at night; season of activity is between May and October; hibernates from late fall to May in fissures in bedrock outcrops and bridge abutments.

REPRODUCTION: mates soon after emerging from the hibernaculum in late April or May; may also mate the preceding fall; female gives birth to 8–12 live young, 18–20 cm (7–8") long, between July and early September.

FOOD: almost exclusively soft-bodied, native crayfish that are growing into their "new" shells.

SIMILAR SPECIES: *Northern Watersnake* (p. 72): lacks low stripe on side; *juvenile:* boldly patterned. *Butler's Gartersnake* (p. 84) and *Common Gartersnake* (p. 94): stripe running down spine. *Red-bellied Snake* (p. 82): red, unstriped belly; smaller.

FRENCH NAME: Couleuvre royale

DID YOU KNOW? The Queen Snake used to be known as the "Leathersnake." It is listed as Threatened by COSEWIC.

DeKay's Brownsnake

Storeria dekayi

Before pollution and massive expansion of housing and transportation infrastructure, DeKay's Brownsnake seemed pre-adapted to city life and was frequently encountered around homesteads. Sadly, much has changed in the past 100 years—this inoffensive snake is found less frequently in urban landscapes than in the past. Still, DeKay's Brownsnake is one of the most frequently encountered snakes in southern Ontario and Quebec. Although not found in the heart of large cities, it continues to be spotted in large parks, green belts and golf courses. As it prefers wet areas, it is often encountered by golfers who spend an inordinate amount of time along water hazards during a frustrating round. • DeKay's Brownsnake is a small, resilient snake, a generalist that does not depend on specific habitat conditions. It feeds on a wide variety of commonly found insects and other invertebrates. It doesn't even require elaborate winter

SIMILAR SPECIES

Common Gartersnake (Eastern subsp.), p. 94

Butler's Gartersnake, p. 84

Eastern Ribbonsnake, p. 92

quarters; rather, the foundation of an old building suits DeKay's Brownsnake perfectly. This tough little snake can spring into action during winter whenever temperatures become unseasonably warm. However, it generally has enough sense to not fully come out of hibernation until the ground thaws in spring.

ID: usually uniform brown; can be grey, yellowish or rusty coloured; 2 rows of small, dark spots run down its back; sometimes parallel spots are linked across spine; wide stripe down spine is faintly lighter than colour of sides; dark, downward stripe near "ear"; belly is light-coloured and generally unmarked. *Juvenile:* light-coloured neck ring.

LENGTH: 30–50 cm (12–20").

DISTRIBUTION: southern Ontario and southern Quebec. *Selected Sites:* Pelee Island, Essex County, Charleston Lake PP, Cottage Country (all in ON).

HABITAT: damp, upland woodlots, freshwater marshes, vacant lots, gardens, golf courses, meadows, urban parks; often under logs, flat rocks or boards.

ACTIVITY PATTERNS: primarily nocturnal; hibernates, often alongside gartersnakes, from fall to spring in old mammal burrows or house foundations.

REPRODUCTION: mates after emerging in spring; female gives birth to 7–15 live young.

FOOD: earthworms, slugs, snails, sow bugs, insects and spiders.

SIMILAR SPECIES: *Common Gartersnake* (p. 94), *Butler's Gartersnake* (p. 84) and *Eastern Ribbonsnake* (p. 92): do not have 2 parallel rows of spots; gartersnake spots, if present, will alternate. *Red-bellied Snake* (p. 82): red belly.

FRENCH NAME: Couleuvre brune

SIMILAR SPECIES

Red-bellied Snake, p. 82

DID YOU KNOW? This snake's name honours James Edward DeKay, a New York naturalist. DeKay's Brownsnake is listed as Not At Risk by COSEWIC, but there are no scientific data on its population abundance or trends, which is the case for all our smaller snakes.

Red-bellied Snake

Storeria occipitomaculata

To us, a Red-bellied Snake is more like a worm with an attitude than a perilous rattler or python. It's such a small serpent that even when completely peeved, all it does is curl its upper lip in agitation. This comical Elvis mimicry has no adaptive advantage when dealing with us, but does serve it well in non-human encounters. Understanding the famed lip curl (or labial protraction) starts with a realization that this "mini-snake" eats earthworms and slugs. The Red-bellied Snake has a small toxin gland near its upper teeth. By smirking while feeding, it not only exposes its relatively long, slug snagging teeth, but also squeezes and empties this gland onto the seized prey. The toxin serves to immobilize the prey. It

SIMILAR SPECIES

Ring-necked Snake, p. 58

DeKay's Brownsnake, p. 80

is very important for this small snake to slow down its prey, for a feisty meal is likely to cause it some harm. The Red-bellied Snake, like all snakes, swallows its prey whole and is not well equipped to deal with something thrashing about. For this reason, it's generally a good idea if the animal is good and dead by the time it reaches the stomach.

ID: brown, grey or black upper body; red, orange or yellow belly; dorsal stripes range from 1 light stripe on back to several dark and light stripes; often 3 light spots on neck.

LENGTH: up to 40 cm (16").

DISTRIBUTION: from eastern Saskatchewan to Nova Scotia but not north of Lake Superior. *Selected Sites:* Ottawa Valley (ON and QC), Cottage Country (ON).

HABITAT: forest edges, meadows and fields; often found under boards, logs or other forest debris.

ACTIVITY PATTERNS: active from April to October.

REPRODUCTION: mates in spring or sometimes in fall; female gives birth to an average of 7 live young in late summer; young mature in 2 years.

FOOD: slugs and earthworms; possibly other invertebrates.

SIMILAR SPECIES: *Ring-necked Snake* (p. 58): much more defined ring around neck; will grow almost twice as long. *DeKay's Brownsnake* (p. 80): has dorsal rows of spots; lacks red belly.

FRENCH NAME: Couleuvre à ventre rouge

DID YOU KNOW? If you manage to find one Red-bellied Snake, keep looking; you are likely to find more. This tiny snake is often quite abundant in pockets of prime habitat.

Butler's Gartersnake

Thamnophis butleri

In Butler's Gartersnake you see that mother nature has a sense of humour—just imagine meeting a wild snake crazily sweeping back and forth and not making any ground. Butler's Gartersnake thankfully tends to use the comedic approach as a defense rather than the unfavourable biting and musking responses used by other gartersnakes. • Butler's Gartersnake has a small distribution, and it is now thought that it once occupied a much larger area in its heyday some 5000 to 7000 years ago. At that time the Great Lakes region had nicely recovered from glaciation, and the climate was warmer and drier. These parameters encouraged prairie-like conditions over much of the Great Lakes region, which were a better fit than this gartersnake's current habitat. • Another factor that has led

SIMILAR SPECIES

Common Gartersnake (Eastern subsp.), p. 94

Eastern Ribbonsnake, p. 92

to its decreased population is a tendency to be a homebody. Studies have shown that unlike other gartersnakes that may range more than a kilometre (0.6 miles) from their hibernaculum, Butler's Gartersnake may move no more than 300 metres (0.2 miles) over the course of a summer. This tendency to stay put makes the snake particularly vulnerable to direct threats because it does not tend to search out new habitat even if the old one has been ruined.

ID: brown with 2 yellow or orange stripes down side and 1 down spine; side stripes cover entire scales on third row up from belly and half the scales on second and fourth rows up; brown sides are marked with dark checkering, mostly along edges of stripes; small head, not much wider than neck.

LENGTH: 30–65 cm (12–26").

DISTRIBUTION: southern Ontario; most common in extreme southwestern Ontario near Windsor and Sarnia, but a few other populations are a little to the north. *Selected Sites:* Luther Marsh, Skunk's Misery, Ojibway Prairie Complex, Windsor and Sarnia airports (all in ON).

HABITAT: remnant tall-grass prairie, wet meadows, pastures, margins of marshes and streams; commonly under boards or rocks.

ACTIVITY PATTERNS: emerges from hibernation in early April; active until mid-September; active during the night, venturing out on warm nights more than other kinds of gartersnakes.

REPRODUCTION: will mate upon emergence in April; 8–10 live young are born in June or July; snakes are 15 cm (6") long at birth.

FOOD: earthworms, slugs, frogs, leeches and salamanders.

SIMILAR SPECIES: *Common Gartersnake* (p. 94): side stripe only covers scales on second and third rows from belly. *Eastern Ribbonsnake* (p. 92): very boldly striped; much slimmer; side stripes along third and fourth rows of scales.

FRENCH NAME: Couleuvre à petit tête

DID YOU KNOW? The Canadian population of Butler's Gartersnake is small, but at some sites this snake can be found in high concentrations. Its global distribution is one of the smallest of any snake species in North America. Butler's Gartersnake is listed as Threatened by COSEWIC.

Terrestrial Gartersnake

Thamnophis elegans

While paddling the silty rivers of the prairies, one may occasionally see a slick brown snake undulate before the bow and swim hastily to the shore. Terrestrial Gartersnakes are not always associated with water but are generally found within travelling distance to it, especially because they like to catch prey both in water and on land. They are likely the most common reptile throughout much of their prairie range but become less conspicuous west of the Rockies. • In fall, Terrestrial Gartersnakes aggregate near mammal burrows or natural crevices before entering hibernation. These harmless snakes frequently make bedfellows with larger species of the prairies by

SIMILAR SPECIES

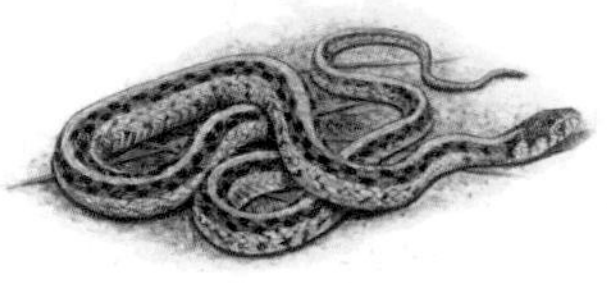

Plains Gartersnake, p. 90

Common Gartersnake (Red-sided subsp.), p. 94

Northwestern Gartersnake, p. 88

denning with Gophersnakes, Prairie Rattlesnakes and Western Rattlesnakes. • The name "Terrestrial Gartersnake" displeases some naturalists because the species is unusually aquatic.

ID: grey-brown; yellow stripe down back; duller yellow stripe down each side; 2 rows of darker spots alternate down back between dorsal and lateral stripes; spots may overlap dorsal stripe, giving it a wavy appearance; few markings on belly, lighter in colour than back and sides; head distinctly wider than neck; in coastal British Columbia, tend to be darker overall, often with a bluish colour.

LENGTH: 45 cm–1 m (18–39").

DISTRIBUTION: from Vancouver Island and the coast of southern British Columbia to southwestern Saskatchewan and north to about the Red Deer River in Alberta. *Selected Sites:* Grasslands NP (SK), Dinosaur PP (AB), Kananaskis Country (AB), Okanagan Valley (BC), southeastern Vancouver Island (BC).

HABITAT: usually found within sight of water; can be found in badlands, mountain valleys, shrublands and woodlands.

ACTIVITY PATTERNS: primarily active during the day, except in extreme cold or heat; hibernate between October and April.

REPRODUCTION: mate upon emerging from the winter den in spring; 4–24 young are born in mid- to late summer.

FOOD: small fishes, frogs, aquatic invertebrates and terrestrial invertebrates.

SIMILAR SPECIES: *Plains Gartersnake* (p. 90): greenish body; orange stripe down back is not wavy; prominent black "whiskers" on upper lip. *Common Gartersnake* (p. 94): on the prairies has red on its sides; in British Columbia has red or black on its head and is generally darker. *Northwestern Gartersnake* (p. 88): more distinct, brightly coloured back stripe; smaller head; white upper lip.

FRENCH NAME: Coulcuvre de l'ouest

DID YOU KNOW? Although it can be argued that the heart of this snake belongs to the arid valleys of the Interior, one can find individuals lurking around tidepools along the rugged coast of British Columbia.

Northwestern Gartersnake

Thamnophis ordinoides

The Northwestern Gartersnake is a loyal resident of the coastal fog belt. It clearly understands that if you live on Canada's west coast, then you should enjoy the rain. This snake not only tolerates the rain, but also purposefully forages during stormy weather, going after moisture-needing earthworms, salamanders and slugs. • It is true that this snake has a fondness for water, but

SIMILAR SPECIES

Terrestial Gartersnake, p. 86

Common Gartersnake (Red-spotted subsp.), p. 94

the love affair seems to end when it comes to swimming. Most gartersnakes take to the water like a Labrador Retriever after a stick, but not the Northwestern Gartersnake. Although it will go for a swim if pushed, this snake seems to get all the moisture it desires from the water that falls from the sky.

ID: can be brown, green, black, red, blue or even albino; may have a white upper lip; almost always a red, orange or yellow dorsal stripe; 2 yellow lateral stripes; yellow or grey belly often has red or black markings; small, blunt head. *Female:* larger; proportionally shorter tail than male.

LENGTH: 45–95 cm (18–37").

DISTRIBUTION: southern coast of British Columbia and all except the northern part of Vancouver Island. *Selected Sites:* Sidney Spit PP, Manning PP (both in BC).

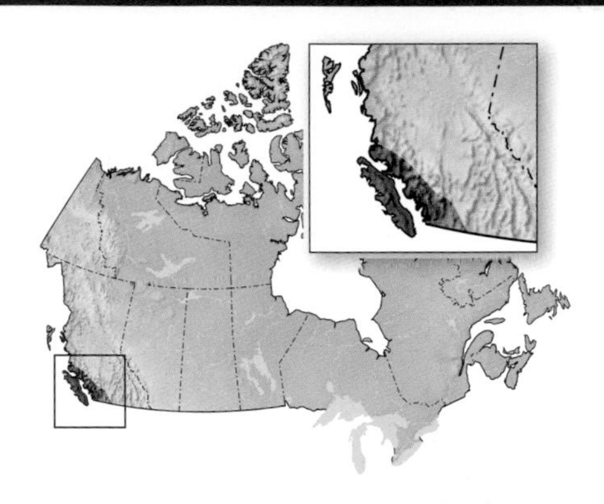

HABITAT: dense bush; open fields; weedy areas.

ACTIVITY PATTERNS: active during hot days; inactive from late fall to early spring.

REPRODUCTION: breeding occurs in either spring or fall; female gives birth to 3–20 live young during summer.

FOOD: invertebrates, fishes, amphibians, small snakes, mammals and birds.

SIMILAR SPECIES: *Terrestrial Gartersnake* (p. 86): black blotches that alternate on either side of a yellow dorsal stripe; dorsal stripe is brown or yellow and wavy because of alternating dark blotches. *Common Gartersnake* (p. 94): never has an orange stripe down back; often quite black; side stripes occur on second and third scale rows.

FRENCH NAME: Couleuvre du nord-ouest

DID YOU KNOW? Other names for the Northwestern Gartersnake include the "Red Racer"—named for the red stripe on most individuals—or the "Garden Snake."

Plains Gartersnake

Thamnophis radix

The Plains Gartersnake is a slough snake. Prairie sloughs are the wetland wildlife factories of the Great Plains, bursting with all forms of wildlife wherever your eyes might fall. The cattails sway with blackbirds, the water is topped with ducks, and the Plains Gartersnake takes its place meandering inconspicuously in the shadows. The subtle prairie hunter ambles on an amphibious course, looking for frogs and sticklebacks under the warm sun of a prairie summer. • The only plain thing about this snake is its name. Beautifully patterned and coloured, this is the finest looking gartersnake on the crowded prairie scene. This colouring, so conspicuous in illustrations and photographs, does not seem to inconvenience the snake it tries to remain hidden in the long prairie grasses from the eyes of hawks and noses of weasels. The alternating dark and light stripes running

SIMILAR SPECIES

Common Gartersnake (Red-sided subsp.), p. 94

Terrestial Gartersnake, p. 86

along the snake mirror the broken shadows that are cast through the stems onto the ground. It is here among shoreline sedges and cattail stalks that the Plains Gartersnake belongs.

ID: dark green to black body; yellow or orange dorsal stripe; yellow and orange dots along each side; little black stripes on upper lip.

LENGTH: 50 cm–1 m (20–39").

DISTRIBUTION: from southern Manitoba to central Alberta north to the Meadow Lake/Cold Lake area. *Selected Sites:* Cold Lake PP (AB), Meadow Lake PP (SK), Miquelon Lake PP (AB), Last Mountain Lake NWR (SK).

HABITAT: found on the plains, "pothole" prairies and southern boreal forests, often near lakes, streams and sloughs.

ACTIVITY PATTERNS: active during the day, except in extreme cold or heat; hibernates in rock crevices, animal burrows or natural sinkholes between October and April.

REPRODUCTION: mates in fall or shortly after emerging from hibernation in spring; female gives birth to 5–30 live young in late summer.

FOOD: small frogs, fishes, aquatic insects and other invertebrates.

SIMILAR SPECIES: *Common Gartersnake* (p. 94) and *Terrestrial Gartersnake* (p. 86): dots or stripes on sides are only on second and third rows of scales.

FRENCH NAME: Couleuvre des plaines

DID YOU KNOW? A male Plains Gartersnake locates a female by detecting chemicals exuded from the female's skin. The male can tell which way she is traveling by comparing intensity of odour on different sides of blades of grass, similar in principle to tracking an animal visually by looking at which way the grass has been bent.

Eastern Ribbonsnake

Thamnophis sauritus

The first characteristics you notice when encountering an Eastern Ribbonsnake are the broad, yellow stripes on the back and sides and the long, tapering tail that accounts for over 25 percent of its body length. The Eastern Ribbonsnake's thin body helps it to swim through favoured ponds and streams, searching for a tasty amphibian. The lack of surface area on the body keeps

SIMILAR SPECIES

Butler's Gartersnake, p. 84

Common Gartersnake (Eastern subsp.), p. 94

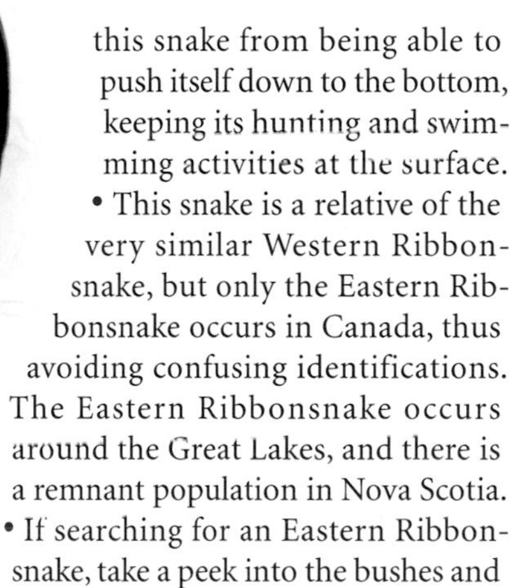

this snake from being able to push itself down to the bottom, keeping its hunting and swimming activities at the surface. • This snake is a relative of the very similar Western Ribbonsnake, but only the Eastern Ribbonsnake occurs in Canada, thus avoiding confusing identifications. The Eastern Ribbonsnake occurs around the Great Lakes, and there is a remnant population in Nova Scotia. • If searching for an Eastern Ribbonsnake, take a peek into the bushes and other vegetation on the fringe of forested wetlands. These snakes like to bask in the sun, but still like to be sure of a quick getaway into the water.

ID: thin body; tail slowly tapers to a point; thick, yellow stripes down sides, 1 stripe down back; stripes end at head; top of head is lighter brown than body; jaw is unmarked and white; distinctive white, crescent-shaped mark in front of eye.

LENGTH: 45 cm–1 m (18–39").

DISTRIBUTION: southern Ontario and southern Nova Scotia. *Selected Sites:* Kejimkujik NP (NS), Rondeau PP (ON), Long Point PP (ON).

HABITAT: edges of forested wetlands, lakes and streams.

ACTIVITY PATTERNS: hunts at night; basks during the day; hibernates during winter.

REPRODUCTION: breeds in spring after emerging from hibernation; female gives birth to 5–20 live young in late summer; young mature in 2–3 years.

FOOD: primarily amphibians such as frogs and salamanders; also some fishes and insects.

SIMILAR SPECIES: *Butler's Gartersnake* (p. 84): thicker body; stripes duller and may be orange on sides; tail less tapered; smaller head; lacks white mark in front of eye. *Common Gartersnake* (p. 94): dark spots on lower jaw; thicker body; tail less tapered; duller stripes; lacks white mark in front of eye.

FRENCH NAME: Couleuvre mince

DID YOU KNOW? There are four subspecies of the Eastern Ribbonsnake, and the Northern subspecies occurs in Canada. The Nova Scotia population is listed as Threatened and the Great Lakes population is listed as Special Concern by COSEWIC.

Common Gartersnake

Thamnophis sirtalis

Red-sided subspecies

The Common Gartersnake is the most widespread reptile on the continent. In every province except Newfoundland, if you have snakes, this one is probably crawling around somewhere. • Canada's best-known snake has somewhat lessened the negative view of snakes in general. In Manitoba in particular, there is a concentration of gartersnakes that is not only a natural phenomenon but also a popular tourist attraction. The Narcisse snake dens in the interlake region of the province may be the largest concentration of any species of reptile in the entire world. Tens of thousands of gartersnakes routinely gather for winter. However, in recent years the number of snakes has declined alarmingly. • Not all aggregations of snakes hold as much favour with people. Warm, insulated basements, a Canadian architectural staple, are perfectly suited to meet the needs of winter-weary snakes. Cracks in the foundations of homes allow snakes to penetrate deep below the frost line and tuck up against a heat-radiating wall. Unfortunately, the occasional snake wiggles its way inside or, more commonly, the masses that

SIMILAR SPECIES

Northwestern Gartersnake, p. 88

Terrestial Gartersnake, p. 86

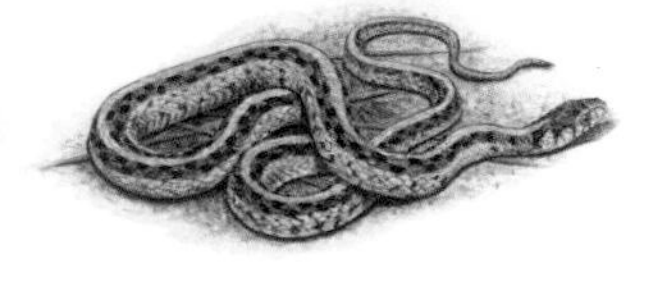

Plains Gartersnake, p. 90

gather on lawns in spring and fall are simply too much of a wilderness experience for most acreage owners. Fortunately, techniques are being developed to move unwanted snakes to naturally occurring hibernating sites.

ID: extremely variable across Canada; colour tends to be dark; 3 yellow stripes down back and sides; side stripes limited to second and third scale rows. *Eastern* subspecies: central Canada; double row of alternating spots between stripes. *Red-sided* subspecies: Alberta to Manitoba; red or orange marks on sides and occasionally on head. *Maritime* subspecies: eastern Canada; brown or grey; poorly defined spinal stripe; tends to have alternating dark marks on back. *Red-spotted* subspecies: British Columbia; black with 3 distinctive, yellow stripes; red spotting along sides and on head.

LENGTH: 85 cm–1 m (33–39").

DISTRIBUTION: from the Pacific to the Atlantic, north to the Alberta–Northwest Territories border; not found in Newfoundland, the grasslands or the Far North. *Selected Sites:* Narcisse (MB), southern Canada in general.

HABITAT: forests, fields and urban areas; most commonly near wetlands.

ACTIVITY PATTERNS: active from April until mid-October; active during the day, but may venture out on warm nights.

REPRODUCTION: breeds usually in spring, but occasionally in fall; female gives birth to 10–30 live young around June; young measure 13–23 cm (5–9"); young reach sexual maturity in 2–3 years.

FOOD: frogs, toads, salamanders, small fishes, mice and earthworms.

SIMILAR SPECIES: *Northwestern Gartersnake* (p. 88): British Columbia coast; most often brown; orange spinal stripe. *Terrestrial Gartersnake* (p. 86): B.C. and Alberta; yellow stripe; dark spots. *Plains Gartersnake* (p. 90): Alberta to Manitoba; orange dorsal stripe; no red spots on sides. *Butler's Gartersnake* (p. 84): southeastern Ontario; smaller; side stripes on second and fourth scale rows. *Eastern Ribbonsnake* (p. 92): southern Ontario and Nova Scotia; very distinct stripes; much slimmer and darker.

FRENCH NAME: Couleuvre rayée

SIMILAR SPECIES

Butler's Gartersnake, p. 84

Eastern Ribbonsnake, p. 92

DID YOU KNOW?
The highest number of visitors to the Narcisse Snake Pits occurs on Mother's Day.

Western Rattlesnake

Crotalus oreganus

Southern British Columbia is the northern limit for the Western Rattlesnake. It is B.C.'s only venomous snake, but despite its reputation as dangerous, it is a calm snake, timid and easily avoided. If a threat approaches, rather than rattling its tail, the Western Rattlesnake prefers to remain still to avoid being seen or heard. However, when it does rattle, the tail vibrations are very fast. Depending on the temperature of the snake (a warm snake will rattle

SIMILAR SPECIES

Gophersnake, p. 76

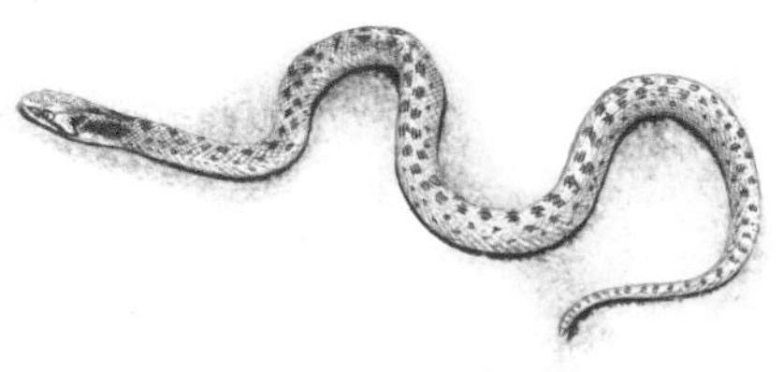

Nightsnake, p. 68

faster than a cool one), it can twitch its tail up to one hundred times per second. • The Western Rattlesnake faces many challenges to survival. Too many snakes are needlessly killed out of fear, while others are unintentionally killed on B.C. highways. In addition, this snake has suffered loss of habitat to human population increases. The Western Rattlesnake has a low rate of reproduction, and only 25 percent of snakes born survive their first year, so when snake populations decrease as a result of human activity, they are slow to recover. • The Western Rattlesnake is legally protected under the British Columbia Wildlife Act.

ID: broad, triangle-shaped head; rattle on end of tail; brown, tan, olive or grey body; large, dark brown blotches along back and smaller blotches along sides; broad, dark stripe on each side of face from eye to jaw; yellowish white, sometimes brownish belly.

LENGTH: 60 cm–1 m (24–39").

DISTRIBUTION: southern interior of British Columbia. *Selected Sites:* Kalamalka Lake PP (BC).

HABITAT: dry, rocky valleys with sparse tree cover for shade and crevices or deep talus slopes for hibernation dens; can be found basking on exposed rock ledges.

ACTIVITY PATTERNS: emerge from hibernation in spring; most follow regular travel routes to favourite foraging and basking areas up to 1 km (0.6 mi) from den; hunt at night; return to same den each fall to hibernate.

REPRODUCTION: mate in late summer, but ovulation and fertilization don't occur until females emerge from hibernation in spring; 2–8 live young born in early fall; males sexually mature at 3–4 years; females sexually mature at 6–8 years; females reproduce only every second or third year.

FOOD: small mammals; usually voles or mice, but the bigger the snake, the bigger the prey.

SIMILAR SPECIES: *Gophersnake* (p. 76): lacks rattle; more regular and distinct pattern; much more white in body. *Nightsnake* (p. 68): lacks rattle; more slender.

FRENCH NAME: Crotale de l'ouest

DID YOU KNOW? The maximum lifespan for the Western Rattlesnake is 25 years. The Western Rattlesnake is listed as Threatened by COSEWIC.

Prairie Rattlesnake

Crotalus viridis

Like cowboys, saloons, tumbleweeds and dust, rattlesnakes just belong to the West. But like those aforementioned symbols of rugged prairie life, rattlesnake lore exceeds rattlesnake truth. Prairie Rattlers are indeed venomous animals, but they pose little threat to humans. Few people have been bitten by a Prairie Rattlesnake, and fewer still have suffered serious injury. Those most vulnerable to the neurotoxins are the young and elderly; to all others, pain and possible loss of limbs would be the

SIMILAR SPECIES

Gophersnake, p. 76

Western Hog-nosed Snake, p. 64

DID YOU KNOW? All rattlesnakes use specialized organs (known as sensory pits) to detect heat. These pits, found between the eye and nostril, allow the snakes to detect their prey in complete darkness. The Prairie Rattlesnake has not yet been assessed by COSEWIC.

extreme outcome of a bite—which is more than enough of a deterrent for stupidity. The rattles on the snakes' tails evolved as an alarm to let heavy-footed, large animals like cattle and bison know where they were to avoid being trampled. Hence, these snakes are very effective at letting humans know where they are. This early warning system allows rattlers to save up their venom for edible fare like mice and voles. It takes a lot of energy to produce the venom; the last thing this rattlesnake wants to do is waste it on a human! • Although Prairie Rattlesnakes can be found on any native prairie pasture or sage flat, they are most commonly encountered around rocky outcrops along coulee bluffs and river valley ridges. These areas are particularly populated during spring and fall, when rattlers migrate to and from communal denning areas. Literally piles of rattlesnakes can be seen sunning themselves in the shallow May or October sun, but come midday summer heat, snakes retreat from the sun, only seeking the heat once again on roads or rocks at night.

ID: large, heavy-bodied; diamond-shaped head; rattle at end of tail; vertical pupils; rough scales above eyes; yellow-green to brown body; dark, irregular blotches down back and sides.

LENGTH: 80 cm–1.4 m (32"–5').

DISTRIBUTION: southern Saskatchewan and southern Alberta. *Selected Sites:* Grasslands NP (SK), Dinosaur PP (AB), Writing-on-Stone PP (AB).

HABITAT: native grasslands, pastures and coulee rims; often found near rock piles or flat boulders; regularly seen on roads at night.

ACTIVITY PATTERNS: active from mid-April to October; tend to be most active in the mornings and evenings that feature bright sunshine; can be found at night on roads.

REPRODUCTION: mating occurs as the snakes emerge from the den in spring; females give birth to 2–8 live young in August or September; reach sexual maturity after 6 years; breed only every 2–3 years.

FOOD: *Adult:* warm-blooded prey such as voles, mice and songbirds; opportunistically prey upon amphibians and reptiles. *Juvenile:* invertebrates.

SIMILAR SPECIES: *Gophersnake* (p. 76): lacks rattle; more regular and distinct pattern; much more white in body. *Western Hog-nosed Snake* (p. 64): lacks rattle; upturned snout; large, dark blotches on belly.

FRENCH NAME: Crotale des steppes

Massasauga

Sistrurus catenatus

As the only rattlesnake to occur east of the prairies, the Massasauga keeps the mythology of the rattler alive and well in a small and decreasing part of Cottage Country. Despite much evidence to soften rattlesnake lore, the Massasauga still suffers from the public fears that have contributed greatly to its decline in the past. Deliberate killings, habitat destruction and the expansion of the road network have forced the Massasauga to reside on the national and provincial endangered species lists. Persistent efforts in Ontario's parks are yielding results, as the Massasauga's image begins to reflect reality. • The Massasauga is the most famous resident of Bruce Peninsula National Park and Killbear Provincial Park, and few people visit these areas without learning of the snake. Encountering the Massasauga, however, is a rare experience. This pleases most visitors to

SIMILAR SPECIES

Eastern Foxsnake, p. 60

Northern Watersnake, p. 72

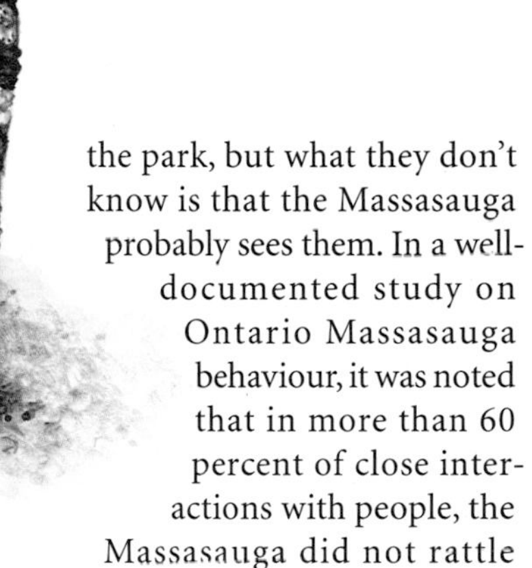

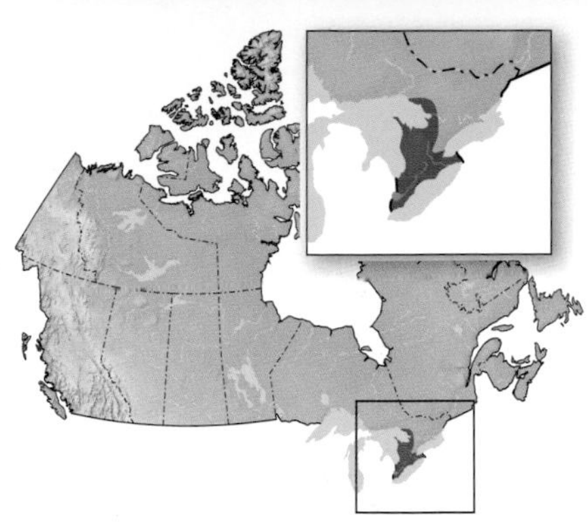

the park, but what they don't know is that the Massasauga probably sees them. In a well-documented study on Ontario Massasauga behaviour, it was noted that in more than 60 percent of close interactions with people, the Massasauga did not rattle or move off. Even when a snake was stepped overtop of, most often the timid snake refused to reveal itself. There was not even a single act of aggression in the 174 recorded incidents. Beneath all the good science there is a fine lesson learned from the snake's inoffensive ways: treat it as it treats you.

ID: brownish-grey body; dark circles on back and sides; rattle at end of tail; vertical pupils; diamond-shaped head; black belly with scattered, light markings.

LENGTH: 60 cm–1 m (24–39").

DISTRIBUTION: Georgian Bay region in Ontario; 2 other small populations persist elsewhere in southern Ontario. *Selected Sites:* Killbear PP, Georgian Bay Islands NP, Bruce Peninsula NP, Ojibway Prairie Complex in Windsor, Wainfleet Bog (all in ON).

HABITAT: associated with water; often near river mouths; may bask on rocks; long grass. The name *Massasauga* in Ojibwa means "great river mouth." This is an appropriate name, as this is a place they commonly occur.

ACTIVITY PATTERNS: hunts actively during the day and during warm evenings; active from late April to October.

REPRODUCTION: breeding occurs in spring after it emerges from hibernation dens; female gives birth to 7–8 live young in late summer; female matures in 3–4 years.

FOOD: mice and voles, frogs, small birds and even other snakes.

SIMILAR SPECIES: *Eastern Foxsnake* (p. 60) and *Northern Watersnake* (p. 72): no rattle on tail; no vertical pupils; head more rounded.

FRENCH NAME: Massasauga

DID YOU KNOW? A newborn Massasauga is cute as a button, with only one piece of rattle on its tail. The Massasauga is listed as Threatened by COSEWIC.

Three of the following four seaturtles should be considered as vagrants. The fourth, the Leatherback, likely moves into coastal Atlantic waters regularly to feed on abundant jellyfish there. Unlike most amphibians and reptiles, these maritime dwellers have the ability to travel tremendous distances in their lives. Whether vagrant seaturtles find themselves lost in Canadian waters in search of food or they follow a warm current north (such as the Gulf Stream), these species, again excepting the Leatherback, are not presently considered regular visitors. These animals tend to prefer warmer oceans and nest on tropical beaches far to the south.

Loggerhead Seaturtle

Caretta caretta

The Loggerhead Seaturtle has a reddish brown shell. The scales on the head and on the top of the flippers are reddish brown, but have yellow borders. The neck, shoulders and upper limbs are brown with some yellow highlights. The underside is also yellow.

LENGTH: *Adult:* 85 cm–1.2 m (33"–4').
FRENCH NAME: Caouane

Green Seaturtle

Chelonia mydas

The Green Seaturtle is not named for its external colour, which just so happens to be green. This turtle is named for its green fat, a rather unusual consequence of its vegetarian diet. It is a large marine turtle with flippers instead of feet. Its shell is hard, unridged and heart shaped, and its face is blunt. The Green Seaturtle is rarely seen in Canada, but when it is seen, it is most frequently encountered off the coast of British Columbia during El Niño years in summer. It may also occasionally occur off the Atlantic provinces.

LENGTH: *Adult:* 75 cm–1.5 m (30"–5').
FRENCH NAME: Tortue verte

Kemp's Ridley Seaturtle

Lepidochelys kempii

Kemp's Ridley Seaturtle is the smallest of the seaturtles in Canadian waters. The grey-olive shell on the adult is almost as wide as long, and the underside is white or yellowish. It has a triangular head and a hooked beak that is well suited to a diet consisting of mainly crabs. Kemp's Ridley Seaturtle is the most endangered of all seaturtles.

LENGTH: 35–75 cm (14–30").
FRENCH NAME: Tortue bâtarde

Leatherback Seaturtle

Dermochelys choriacea

Also commonly known as the "Luth," the endangered Leatherback Seaturtle is perhaps the most original of all seaturtles. Its origins appear to date most of the way back to the dinosaurs, and it is genetically distinct from the other seaturtles of the world. This table-sized ocean leviathan has flippers instead of feet. Its brown shell is not hard plated, and there are seven ridges running lengthwise. The Leatherback roams the world's oceans, encountering both icebergs and palm trees in its search for jellyfish. It is this delicacy that routinely draws the Leatherback to northern areas of both coasts of Canada and even to Baffin Island in the Arctic.

LENGTH: 1.3–2.4 m (4–8').

FRENCH NAME: Tortue luth

Red-eared Slider

Trachemys scripta

At one time, pet stores all across Canada sold small turtle hatchlings. Millions of these cute and inexpensive reptiles found their way into so many homes that the Red-eared Slider claimed the distinction of being the most common reptilian pet in North America and the world. Recently, some provincial jurisdictions have outlawed the sale of these pets, but they nevertheless remain common in the homes of Canadians. • The principal ecological problem of Red-eared Sliders is not that they are captive pets (although captives can harbour salmonella bacteria), but rather that the cute hatchlings grow too large for their owners and are released to the wild. The milder regions of the Maritimes, Quebec, Ontario and British Columbia can and do support individual Red-eared Sliders. These animals compete with native turtles for limited ecological resources and can transmit parasites or diseases to wild, naturally-occurring turtles. People wishing to get rid of their pet Sliders should find new owners for them or contact one of many organizations that will place their turtles in happy homes.

ID: looks similar to the *Painted Turtle* (p. 32); characteristic red (occasionally yellow) broad band behind eye.

LENGTH: 13–29 cm (5–11 1/2").

FRENCH NAME: Tortue à oreilles rouges

European Wall Lizard

Podarcis muralis

In name, architecture and culture, Victoria, British Columbia, has a distinctly European quality. The less conspicuous but no less significant European Wall Lizards are the most recent European import. Currently, there are three known populations of European Wall Lizards on the Saanich Peninsula of Vancouver Island, but when or how the lizards first arrived remains mysterious. The colonization appears to be a recent event: biologists tracking population changes have noted healthy rates of range expansion. Elsewhere in North America, populations of these wall dancers were established through reluctant pioneer lizards carried from Europe, or by escaped pets. In Victoria initial sightings of European Wall Lizards were made near stables, suggesting that the "pioneers" may have arrived in hay shipments. What effects European Wall Lizards will have on Victoria's natural community are not yet known. Regardless of how or when they arrived, these lizards are apparently satisfied with their new home.

ID: distinct neck with a collar of large scales on throat; skinny body, slender, slightly flattened form, even through head; slender tail; feet have long and slender toes; colour and patterns are quite variable among and within individuals but generally have a scaly look; tongue is long and forked.

LENGTH: 15–23 cm (6–9"); females are slightly smaller than males.

FRENCH NAME: Lézard des murailles

Pacific Pond Turtle

Actinemys marmorata

This reluctantly social animal basks on emergent logs in close proximity to others—sometimes too close—but nips and off-balance tussels end quickly among these armoured gladiators, and peace soon returns to the sun worshippers. In Canada, any sighting of a Pacific Pond Turtle would be considered exceptional, as this West Coast reptile has unfortunately long disappeared from our ponds. The last confirmed sighting dates to the 1960s, while the only two specimens that conclusively tie these turtles to Canada were collected in the 1930s. • There is some debate over whether or not the Pacific Pond Turtles in British Columbia represented a native population or whether people introduced them to the province. What is known is that this Canadian turtle was eaten in large numbers not by some dreaded introduced predator, but by humans. This species is quite wary, perhaps as a result of relentless hunting of it for the fish markets of Vancouver and Seattle. Its numbers and distribution have declined drastically throughout its range, but recent recovery efforts in Washington and Oregon have been quite successful.

ID: variable colour, from dark brown to olive to black; yellow underside; shell is relatively flat and patterned with dark spots; dark spots on head and legs.

LENGTH: 12–14 cm ($4^1/_2$–$5^1/_2$").

FRENCH NAME: Tortue des etang de l'ouest

Eastern Box Turtle

Terrapene carolina

Eastern Box Turtles are a common animal of the reptilian pet trade. Although they are docile, easy to care for and grow no larger than a Kaiser roll, they do live a long time. Box Turtles over 120 years are known in the world, and it is often this longevity that makes them candidates to be released into the wild and, more unfortunately, candidates for extinction. Although there is some discussion as to whether or not Eastern Box Turtles are native to Ontario, they are found in the eastern half of the United States and around the southern Great Lakes. There is solid evidence that Eastern Box Turtles occurred in southern Ontario up to about 1600. Their remains have been found in archeological sites in the garbage pits of Native people. The turtles were eaten and their shells were used for ornament and as a form of currency. It is thought that harvesting adults of these long-lived turtles led to their extirpation from parts of their northern distribution in the United States and southern Ontario. Specimens found today in these regions are likely released captives. Nevertheless, the practice of releasing any non-native animal into a foreign community is an ill-advised activity, and too often leads to trouble for the naturally occurring critters.

ID: stocky body and limbs; brownish or black, highly-domed shell with variable orange, red or yellow markings; distinct hinge across middle of plastron, allowing turtle to be completely enclosed in its shell; feet webbed at base of toes only. *Male:* red eyes. *Juvenile:* more intense colouring.

LENGTH: 12–19 cm ($4^1/_2$–$7^1/_2$").

FRENCH NAME: Tortue tabatière

Pigmy Short-horned Lizard

Phrynosoma douglasii

Limited habitat and human disturbance might have been the reason for the extirpation of the Pigmy Short-horned Lizard from the sun-baked cliffs of Osoyoos, British Columbia. There have been no confirmed Canadian reports for almost a century. Unlike other types of lizards found in the Okanagan valley, the Pigmy Short-horned Lizard is not hyperactive or a high-speed risk taker. It is content to bask obscurely, relying on cryptic colouration rather than speed to protect it from a hungry world. • The female Pigmy Short-horned Lizard is considerably larger than the male, likely to allow room for her developing young that are born alive. Live birth of the young enables the mother to provide greater control over the temperature of the embryos. This likely allows this lizard to exist so far north. Until recently, this species was considered to be a subspecies of the Short-horned Lizard. Scientists have now concluded that the two animals are separate and distinct species (see Greater Short-horned Lizard, p. 42).

ID: rounded body; wide base to tail that quickly comes to a point; cryptic colouration covers body—can be grey, brown, yellow or rust-coloured; rough, pointed skin; cream-coloured belly.

LENGTH: 6–8 cm (2½–3").

FRENCH NAME: Iguane à petites cornes de Douglas

Timber Rattlesnake

Crotalus horridus

To those who mourn the lost of wildness, the absence of one of the world's great snakes gives pause for thought. The Timber Rattlesnake was a convenient target for decades by persons scared of or ill-informed about snakes. The persecutors of Canada's Timber Rattlesnake reached their ultimate goal by 1941. Previous to that last Canadian record, parties would radiate across southern Ontario and kill every Timber Rattler encountered. Slaughtering reached its climax at communal overwintering dens, where snakes would gather by the dozens to escape winter's chill. There is no other form of wildlife in Canada that has been so callously eliminated. Our social and environmental ethics have evolved greatly since the Second World War, but habitat alteration combined with rattlesnake perception likely hinders all thoughts of this animal returning to Canada. If nothing else, the memory of the Canadian Timber Rattlesnake provides a reference point in the growth of our collective environmental ethic, which has happily evolved since the dark days of snake hunts.

ID: rattle on end of tail; base colour may be yellow, brown, grey or black; dark blotches on front half of body; bands on rear of body; vertical pupils.

LENGTH: 90 cm–1.5 m (3–5'); up to 1.8 m (6').

FRENCH NAME: Crotale des bois

Amphibians

Eastern Newt

Notophthalmus viridescens

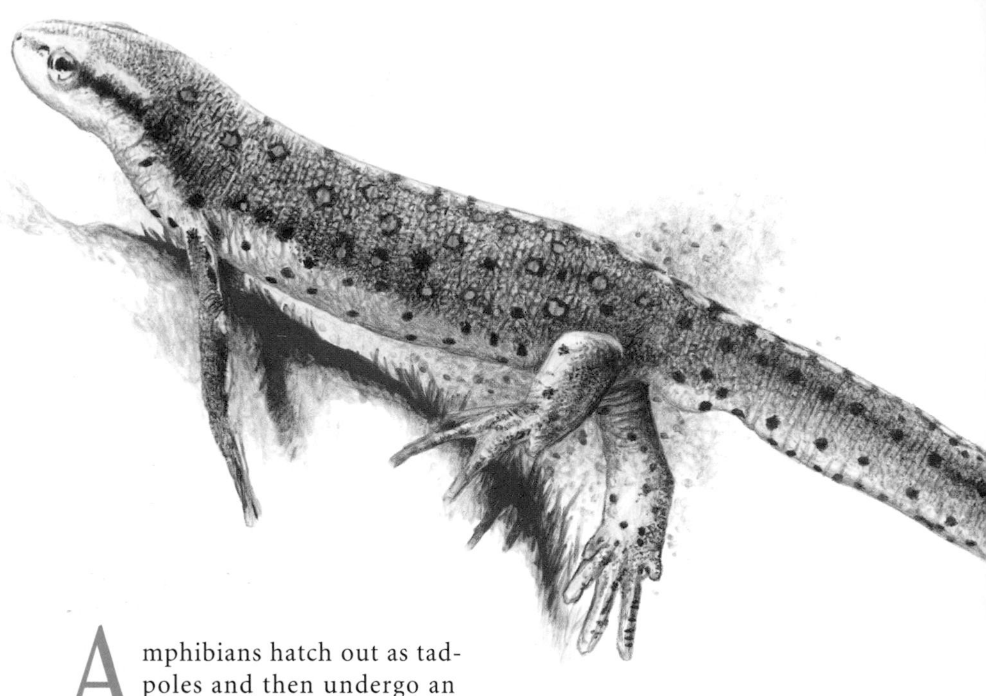

Amphibians hatch out as tadpoles and then undergo an elaborate morphing process wherein legs and lungs are grown and gills are absorbed. The Eastern Newt experiences life in the egg, life as a tadpole and life as an adult like most amphibians, but newts alone in Canada also experience life as an "eft." The eft stage follows the tadpole stage, and it is a bit like a rebellious teenage phase for the newt. It wanders alone through forests decked out in vibrant colours, armed with the knowledge that just about all other critters are wise enough to leave it alone. The vibrancy of its red skin announces that it is toxic to any creature that attempts to bite it. The eft stage lasts for three to four years before its final transformation into the adult form. This change, just as radical as the previous change from a tadpole, has the eft losing its terrestrial adaptation in favour of its upcoming aquatic needs. The tail flattens, and the skin turns almost

SIMILAR SPECIES

Spotted Salamander, p. 126

DID YOU KNOW? The Eastern Newt has a neurotoxin in its skin that is lethal to many predators. The intense colour of the eft, combined with the display postures that a threatened animal contorts into, warns would-be predators that it is toxic.

completely green save for a few red spots that hint back to the eft stage. It is only when it is fully transformed into a newt that it can mate and reproduce. The wonderful life cycle of the Eastern Newt thus begins and ends in the water, but in between, a noteworthy experience is had on land.

ID: *Adult:* light brown to greenish body; black-bordered, orange-red spots along sides; small, black specks scattered all over body; yellow belly; flattened, fin-like tail. *Eft:* bright red, although individuals nearing sexual maturity are less colourful; rough skin; black-rimmed, red spots on sides; tiny, black specks scattered all over body.

LENGTH: *Adult:* 7–14 cm (3–5$^{1}/_{2}$"). *Eft:* 4–8 cm (1$^{1}/_{2}$–3").

DISTRIBUTION: from western Ontario to Cape Breton Island, but absent north of Lake Superior. *Selected Sites:* anywhere within its range and habitat.

HABITAT: *Adult:* ponds, swamps, slow-moving streams and lakes. *Eft:* moist leaf litter or under logs in forests.

ACTIVITY PATTERNS: efts are seen more often during the day than during the night; some adults and all efts hibernate on land from late fall to early spring; other adults remain in the water over winter.

REPRODUCTION: following involved spring courtship displays, female lays 200–400 eggs singly on submerged vegetation; tadpoles emerge after 1–2 months; tadpoles transform into efts by the end of summer; remain as efts for 3–4 years, after which they transform into aquatic, sexually mature adults and most likely return to their home pond; can live for more than 13 years.

FOOD: *Adult:* aquatic organisms such as insects, small crustaceans, mollusks, amphibian eggs and larvae. *Eft:* invertebrates including springtails, mites and spiders.

SIMILAR SPECIES: adult Eastern Newt may be confused with a young *Spotted Salamander* (p. 126), which lacks red spots.

FRENCH NAME: Triton vert

Rough-skinned Newt

Taricha granulosa

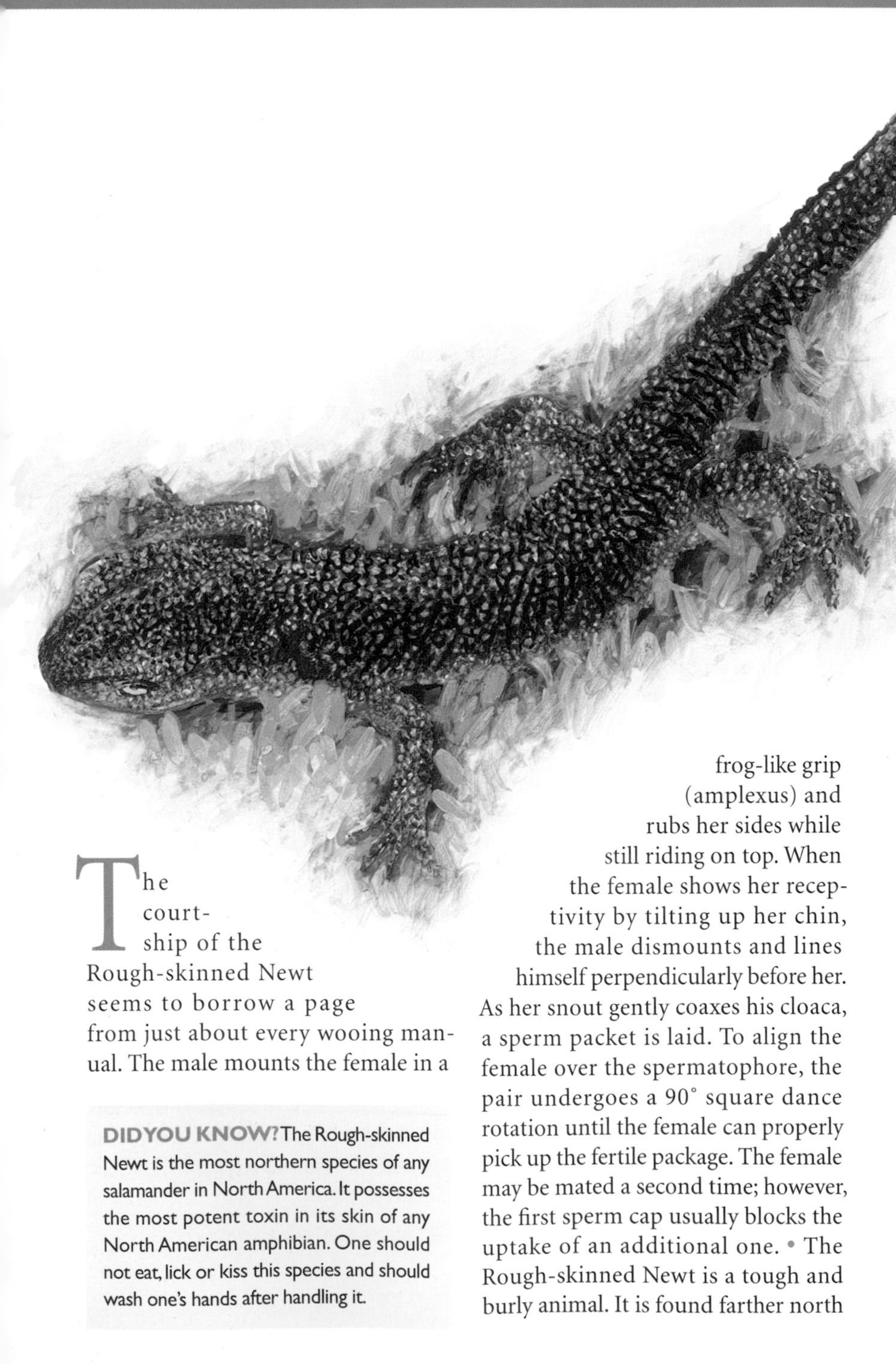

The courtship of the Rough-skinned Newt seems to borrow a page from just about every wooing manual. The male mounts the female in a frog-like grip (amplexus) and rubs her sides while still riding on top. When the female shows her receptivity by tilting up her chin, the male dismounts and lines himself perpendicularly before her. As her snout gently coaxes his cloaca, a sperm packet is laid. To align the female over the spermatophore, the pair undergoes a 90° square dance rotation until the female can properly pick up the fertile package. The female may be mated a second time; however, the first sperm cap usually blocks the uptake of an additional one. • The Rough-skinned Newt is a tough and burly animal. It is found farther north

DIDYOU KNOW? The Rough-skinned Newt is the most northern species of any salamander in North America. It possesses the most potent toxin in its skin of any North American amphibian. One should not eat, lick or kiss this species and should wash one's hands after handling it.

than any other salamander in North America, even reaching the Alaskan frontier. When threatened, the newt will display its bright belly, warning off predators. Perhaps the greatest natural threat to the Rough-skinned Newt is the Common Gartersnake, which is more resistant to the newt's tetrodotoxin than other snakes on the west coast. The newts from Vancouver Island were found to lack tetrodotoxins. Not surprisingly, the gartersnakes on the island show a corresponding decrease in resistance to the toxin.

ID: warty, grainy skin; light to dark brown on top; yellow to orange belly; yellow eyes.

LENGTH: 12–20 cm (5–8").

DISTRIBUTION: entire coast and islands of British Columbia. *Selected Sites:* Goldstream PP, Newcastle Island PP, Pacific Rim NP (all in BC).

HABITAT: *Adult:* ponds, lakes and slow tributaries. *Eft:* under logs, bark and other debris in forests; occasionally fields and pastures.

ACTIVITY PATTERNS: adult tends to move at night, but eft can be quite commonly seen during the day; both overwinter on land from early October to March or April.

REPRODUCTION: elaborate courtship display prior to breeding in spring; female lays her eggs singly on submerged plants; eggs hatch in 3–4 weeks; by late summer or occasionally the following summer, the tadpoles transform into efts; the eft stage can last up to 5 years until the eft becomes a sexually mature adult (newt).

FOOD: amphipods, both aquatic and terrestrial insects, leeches, tadpoles, mollusks and worms.

SIMILAR SPECIES: none.

FRENCH NAME: Triton rugueux

Mudpuppy

Necturus maculosus

The Mudpuppy is an aquatic salamander equipped with full-time gills that spends its life on the bottom of slow-moving streams and lakes. It is very seldom seen except by the angler who unexpectedly pulls one up from the depths. The Mudpuppy, with an alien twinkle in its small eyes, undoubtedly shocks and surprises those anglers who see it for the first time. The surprise is more extreme when the Mudpuppy becomes an ice-fisher's bycatch, for this salamander is active year round. Those sportsmen uneducated with the world of wildlife too often panic and kill this unusual but harmless amphibian on sight. To those more educated, the hook is carefully removed and the Mudpuppy is gently returned to the water. • The Mudpuppy is much more than a disappointing catch for an angler. As a long-lived member of Canada's aquatic community, it serves as an indication of the state of our waters. It tends to occur in unpolluted waters because its skin easily absorbs whatever is dissolved in the water. Unfortunately, the Mudpuppy's range in Canada closely

SIMILAR SPECIES

neotenic Tiger Salamander, p. 130

mirrors the growth of intense industry, and the health of the Mudpuppy in its region reflects this overlap. Along a stretch of the St. Lawrence River that showed the highest levels of PCBs in the water, over 60 percent of the Mudpuppies examined had some type of deformity. Although the Mudpuppy does normally look odd, an unusual number of toes or limbs caused by industrial pollutants is not very charming.

ID: long, feathery, dark red external gills; grey to rusty brown body with dark blue spots on back; grey belly; flattened, fin-like tail; 4 toes on both the front and hind feet; dark eye line runs through eyes to gills. *Male:* papillae around cloaca. *Female:* light-coloured region surrounding cloaca.

LENGTH: 20–45 cm (8–18").

DISTRIBUTION: southeastern Manitoba to the region of Quebec City; not found north of Lake Superior. *Selected Sites:* Grand River, Thames River, Sydenham River, southern Great Lakes (all in ON).

HABITAT: lakes, rivers and streams; permanent water that does not freeze to the bottom in the winter is vital.

ACTIVITY PATTERNS: completely aquatic; active primarily after dark; during the day often found under rocks, but also found in weedy or muddy waters; less active during winter.

REPRODUCTION: mates in spring; female builds a nest under a rock or a board and deposits her eggs singly on the underside surface; female guards the eggs until they hatch; young take 4–6 years to reach sexual maturity; may live up to 30 years.

FOOD: crustaceans, insects, fishes and worms.

SIMILAR SPECIES: *neotenic Tiger Salamander* (p. 130): more rounded snout; 5 toes on hind feet; more prominent eyes; range overlaps only in Manitoba.

FRENCH NAME: Necture tacheté

DID YOU KNOW? The Mudpuppy is one of the only animals in this book that could be found to be active on any day of the year.

Northwestern Salamander

Ambystoma gracile

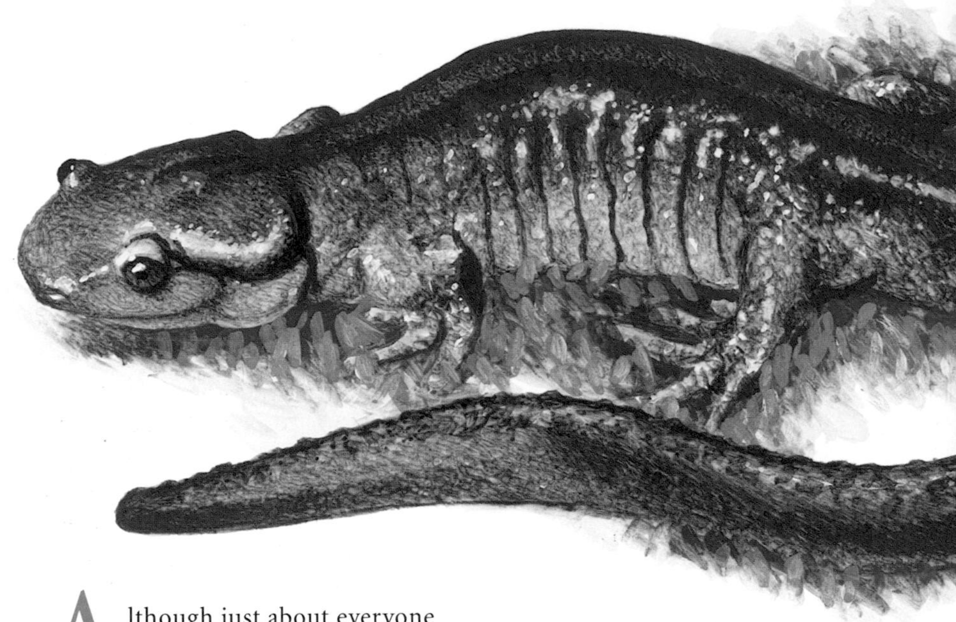

Although just about everyone along Canada's west coast is apt to complain at some point of the never-ending winter rains, if the Northwestern Salamander could have a say in the matter, it wouldn't change a thing. When the rains turn the land soggy, this usually secretive salamander ambles into forest openings to enjoy the shower. Follow it long enough and it will lead you to a pond or, more commonly, a hideout into which it disappears until the rains encourage another journey. • As successful as salamanders have been over time, vigilance is not one of their highpoints. The Northwestern Salamander, like other mole salamanders, is not a speedster, nor is it particularly aware of threats when plodding around. Such a good-sized salamander would make a desirable meal for just about any hungry beast; therefore, the Northwestern Salamander has a trick up its sleeve—or along its back.

SIMILAR SPECIES

Long-toed Salamander, p. 124

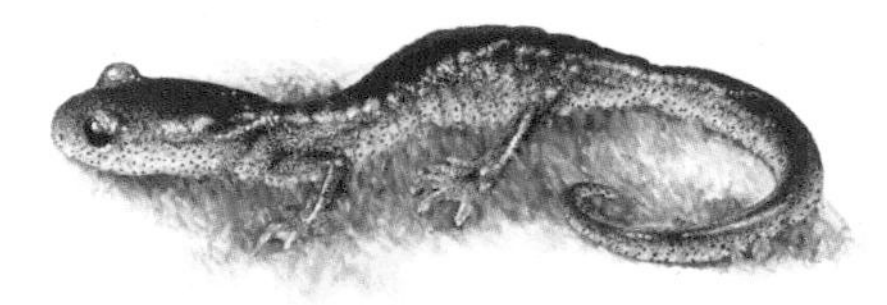

Ensatina, p. 140

Prominent glands are found on the neck and along the ridge of the tail of this salamander. These glands produce a mild toxin that turns the stomach of would-be predators. Even while gently handling an individual caught during a rainstorm, white, sticky ooze will coat your fingers. This is not to say that a Northwestern Salamander is immune to predation. It is certainly fed upon by a gamut of predators who tolerate the toxin, but likely suffer at some level from it.

ID: grey-brown body; beige belly; large, lighter-coloured glands located right behind eyes and along top of tail. *Neotenous:* does not have evident glands.

LENGTH: 14–22 cm (5½–8½").

DISTRIBUTION: British Columbia from the islands east to the coast and Cascade mountain ranges. *Selected Sites:* Marion Lake (BC).

HABITAT: humid areas; typically in dense forests but also in meadows and fields.

ACTIVITY PATTERNS: can be active during or after rain, particularly after dark; inactive during the coldest time of the year.

REPRODUCTION: breeds in marshes, lakes or slow-moving streams from April to July depending on elevation and temperature; once the female has received the spermatophore from the male, she will lay 6–12 small egg masses of 15–35 eggs; eggs hatch within 4 weeks; tadpoles will transform into adults in 1–2 years; some individuals remain aquatic as adults (neotenous).

FOOD: *Adult:* aquatic or terrestrial invertebrates. *Tadpole:* plants, invertebrates.

SIMILAR SPECIES: *Long-toed Salamander* (p. 124): light stripe all along back; lacks prominent neck glands. *Ensatina* (p. 140): lacks neck glands; constricted tail base.

FRENCH NAME: Salamandre foncée

DID YOU KNOW? The Northwestern Salamander was known, until recently, as the "Brown Salamander." The Northwestern Salamander is unusual in that males produce clicking sounds when confronted by predators and also during mating.

Jefferson Salamander

Ambystoma jeffersonianum

Long toes and a pointy nose set the Jefferson Salamander apart from the other members of the mole salamander family. This salamander's range extends into Canada from the Carolinian forests of the United States. Lake Erie shores and the Oak Ridges Moraine have at least 22 known sites with populations of the Jefferson Salamander. There could be more populations, but finding them is tricky; the Jefferson Salamander spends most of its time underground, and it is nocturnal, so on the rare occasion it ventures outside, it is usually dark. • The only time that this reclusive amphibian could easily be seen is during its annual migration to breeding ponds. For most salamanders, this migration is one of the most dangerous times in their lives; cars, predators and a lack of moisture on the trek are all factors that could lead to their demise. To avoid drying out, the Jefferson Salamander crawls out of its burrow and heads for the ponds right after the snow has melted. • A threat to the presence of the Jefferson Salamander is the intensity of devel-

SIMILAR SPECIES

Blue-spotted Salamander, p. 122

opment in the golden horseshoe region of Ontario. Habitat destruction is a significant threat to this species. The Jefferson Salamander, like most other pond-breeding amphibians, is susceptible to predation by fish and usually disappears from ponds that have permanent fish populations.

ID: brown-grey body with light blue-grey sides; younger adults may have bluish spots on their sides; very long toes; pointed snout; tail as long as body.

LENGTH: 10–21 cm (4–8").

DISTRIBUTION: Norfolk County in southern Ontario, Niagara Escarpment and Oak Ridges Moraine. *Selected Sites:* Bachus Woods, Hilton Falls Conservation Area, Bruce Trail on Niagara Escarpment (all in ON).

HABITAT: *Breeding:* fish-free small ponds; agricultural ponds with aquatic vegetation and little pesticide residue. *Non-breeding:* older, moist, deciduous forests; in burrows or underneath logs and large leaf litter piles.

ACTIVITY PATTERNS: occasionally active on the surface during rainfall; primarily remains underground or under leaf litter.

REPRODUCTION: migrates to ponds during snow melt (usually March); female will internally fertilize eggs and lay them singly or in clumps of 150–300 on stems of aquatic vegetation; young hatch after 21 days and take 2–4 months (usually by August) to grow into adults.

FOOD: many invertebrates from the soil; earthworms, beetles, ants.

SIMILAR SPECIES: *Blue-spotted Salamander* (p. 122): blue spots on sides, legs and tail; shorter toes; more rounded snout; smaller.

FRENCH NAME: Salamandre de Jefferson

DID YOU KNOW? The Jefferson Salamander migrates farther than most other salamanders to breed, and this makes it especially vulnerable to mortality on roads. The Jefferson Salamander is listed as Threatened by COSEWIC.

Blue-spotted Salamander

Ambystoma laterale

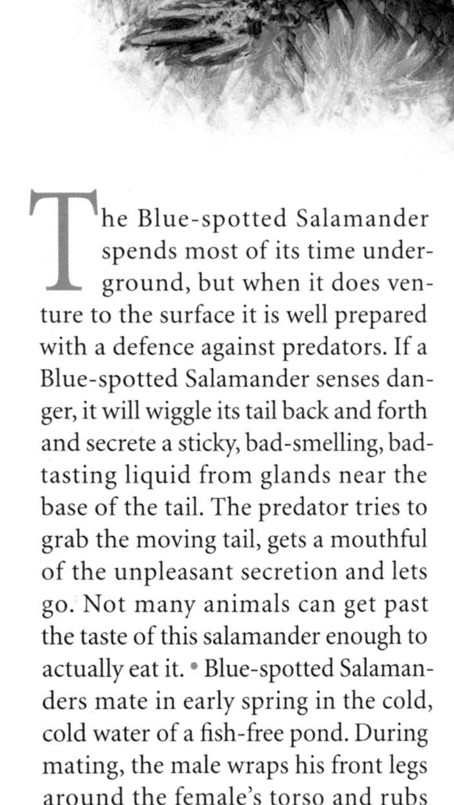

The Blue-spotted Salamander spends most of its time underground, but when it does venture to the surface it is well prepared with a defence against predators. If a Blue-spotted Salamander senses danger, it will wiggle its tail back and forth and secrete a sticky, bad-smelling, bad-tasting liquid from glands near the base of the tail. The predator tries to grab the moving tail, gets a mouthful of the unpleasant secretion and lets go. Not many animals can get past the taste of this salamander enough to actually eat it. • Blue-spotted Salamanders mate in early spring in the cold, cold water of a fish-free pond. During mating, the male wraps his front legs around the female's torso and rubs his snout on the back of her head. After a few hours of excitement, the male releases the female and walks in front of her slowly. She nuzzles his cloaca and he deposits a spermatophore on a leaf or branch on the bottom of the pond. The female then picks up the spermatophore with her cloaca and fertilizes her eggs. • Because the Blue-spotted Salamander spends most of its time underground, the best time to see it is in spring, when it migrates to breeding ponds, or during rainy nights in late summer, when newly transformed young migrate to their terrestial habitat. On one April night in Nova Scotia, over 100 salamanders were seen on their way to the breeding ponds.

SIMILAR SPECIES

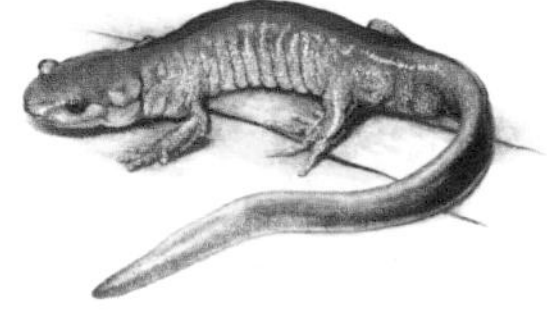

Jefferson Salamander, p. 120

ID: dark blue overall, with large, blue, irregular spots primarily on upper surface. *Female:* larger, with relatively shorter tail. *Juvenile:* brighter blue spots.

LENGTH: 7–13 cm (3–5").

DISTRIBUTION: from southeastern Manitoba to Nova Scotia and Labrador. *Selected Sites:* Hilton Falls Conservation Area, Bachus Woods (both in ON).

HABITAT: most native habitats including mixed woods or continuous forest; may breed in temporary wetlands or lakes.

ACTIVITY PATTERNS: aside from the breeding season, adults are terrestrial, living underground in burrows; can be found moving around during and following rain, particularly at night.

REPRODUCTION: a few days after the female receives the spermatophore, she will lay up to 200 eggs in the water; eggs are attached to vegetation or rocks on the bottom as singles or in small, loose clumps; tadpoles transform in late summer.

FOOD: insects, other invertebrates and even tadpoles of other amphibians.

SIMILAR SPECIES: *Jefferson Salamander* (p. 120): light-coloured marks confined to limbs and sides; lighter base colour; lacks a dark belly; ranges overlap only in southern Ontario.

FRENCH NAME: Salamandre à points bleus

DID YOU KNOW? In the past, Blue-spotted and Jefferson salamanders hybridized and created two all-female triploid species: Tremblay's Salamander and Silvery Salamander. Tremblay's females breed with Blue-spotted males, and Silvery females breed with Jefferson males.

Long-toed Salamander

Ambystoma macrodactylum

Finding these small, handsome recluses is not easy, which makes their discovery even more exciting. Like all mole salamanders in Canada, the Long-toed Salamander moves to the wetlands in spring to breed and back to woodland cover by summer. Winter is spent hibernating about 50 to 70 centimetres (20 to 28 inches) below the ground, sometimes in groups of up to 14 salamanders. • A breeding population of Long-toed Salamanders was recently discovered on the edge of the Long-toed Salamander's Canadian range in Waterton Lakes National Park. The breeding pond was located across the road and down from the park's visitor centre, just a few hundred metres from Waterton's landmark hotel. During the springtime breeding crunch, the salamanders were discovered crossing the road, only to be trapped on the asphalt by a newly constructed squared curb. This 15-centimetre (6-inch) "cliff" barred access to the breeding pond and would have eventually threatened the population. Park

SIMILAR SPECIES

Western Red-backed Salamander, p. 152

Coeur d'Alene Salamander, p. 148

officials promptly replaced the obtrusive curb with one that is sloped and salamander friendly.

ID: orange-yellow stripe along back from head to tail; dark background colour, typically black but occasionally dark green; may be some orange on upper legs; small, light-coloured highlights on sides and legs; second toe from outside on hind feet is longer than other toes.

LENGTH: 10–17 cm (4–6½").

DISTRIBUTION: from Vancouver Island and the British Columbia coast to the foothills of Alberta; as far north as the Stikine and Peace rivers. *Selected Sites:* Manning PP (BC), Waterton Lakes NP (AB), Banff NP (AB).

HABITAT: moist mountain, foothill and coastal forests; a water body should be nearby, preferably one without fish.

ACTIVITY PATTERNS: active at night during late spring and early summer and during rainfall; seasonally active in early spring and fall; at other times of the year, tends to remain inactive under logs or rocks or underground in rodent burrows.

REPRODUCTION: one of the earliest breeders in Canada; snow melt likely triggers breeding; female lays about 60 eggs singly or in small egg masses on vegetation or rocks; eggs hatch within 2–5 weeks; tadpoles leave the breeding ponds by the end of summer; in some alpine areas, larval development may take 2 years.

FOOD: insects and other invertebrates.

SIMILAR SPECIES: *Western Red-backed Salamander* (p. 152): red or orange stripe on back; more slender. *Coeur d'Alene Salamander* (p. 148): restricted to the Creston, B.C. area; nasal grooves and shorter toes.

FRENCH NAME: Salamandre à longs doigts

DID YOU KNOW? There are five subspecies of the Long-toed Salamander. The Eastern, Western and Northern subspecies all occur in Canada, whereas the Southern and Santa Cruz subspecies occur only in the United States. The farther south the population, the more likely the yellow stripe on the salamander's back will be ragged-edged or discontinuous.

Spotted Salamander
Ambystoma maculatum

The animal kingdom contributes to the earthly explosion of spring with the impressive form of the Spotted Salamander. Once this polka-dotted "plough" has dug its way to the surface, the Spotted Salamander wanders to ponds to breed, finding refuge during the day from the spring sun under logs or other debris. It eventually finds a suitable wetland and cruises the bottom in search of another spotted suitor to mate. • Breeding time is the best opportunity to observe this salamander, as the clear and shallow breeding ponds allow a window of observation. Its surprising colour and light, astronaut-like walk along the bottom are a fantastic spring sight. The fun for salamander-watchers is over too quickly, as the business of mating and egg-laying is done with great speed. By the time the last of the songbirds have arrived to establish territories, the Spotted Salamander adult may already be wandering away from the breeding ponds to find shelter and

SIMILAR SPECIES

Jefferson Salamander, p. 120 Blue-spotted Salamander, p. 122 Eastern Newt, p. 112

food under debris and loose soils. From that time of year on, this digger is hard to find. To appease the soft spots many herpetologists have for the Spotted Salamander, spring alone is the time for admiration.

ID: blue-black body with large, yellow or orange spots throughout; grey belly.

LENGTH: 15–25 cm (6–10").

DISTRIBUTION: from Lake Superior east to the Maritimes. *Selected Sites:* Fundy NP (NB), around ponds or small lakes throughout its range.

HABITAT: forested areas surrounding ponds.

ACTIVITY PATTERNS: active mostly at night from early spring to early fall; spends most of its time up to 1 m (39") underground in rodent burrows.

REPRODUCTION: breeding occurs in early spring; breeding pond is shallow and fishless; female can lay up to 250 eggs in clumps or somtimes in a single mass that is attached to underwater vegetation; eggs hatch in 1–2 months; tadpoles transform in another 2–3 months; males mature in 2–3 years, females in 3–5 years.

FOOD: primarily carnivorous throughout its lifecycle; invertebrates.

SIMILAR SPECIES: *Jefferson Salamander* (p. 120) and *Blue-spotted Salamander* (p. 122): no yellow spots. *Eastern Newt* (p. 112): yellow belly; orange-red spots.

FRENCH NAME: Salamandre maculée

DID YOU KNOW? The bright spots on this salamander presumably serve as a warning to potential predators. Like most salamanders, the Spotted Salamander has toxins in its skin that discourage predators from eating it.

Small-mouthed Salamander

Ambystoma texanum

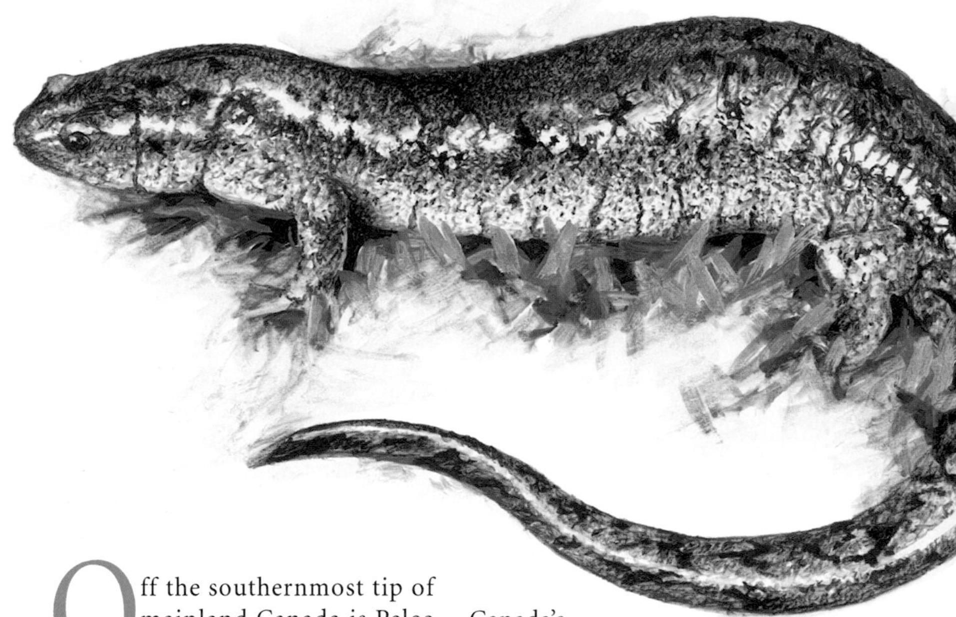

Off the southernmost tip of mainland Canada is Pelee Island, famous for migratory songbirds and the only toehold in Canada for the Small-mouthed Salamander. Although still common on its island refuge in Canada, this animal has a much wider North American distribution. As suggested by its species name *texanum*, this salamander can be found under logs and rocks as far south as Texas and Louisiana, while also holding the current title of Canada's southernmost salamander resident. • Like so many others in its family, the Small-mouthed Salamander can best be described as a recluse based on its activity (or lack thereof) for much of the year. This perennial pattern of inactivity is broken for just a few weeks by the first rains of spring. As water seeps underground, the Small-mouthed Salamander awakens, drawn

SIMILAR SPECIES

Blue-spotted Salamander, p. 122

Spotted Salamander p. 126

to breeding ponds by hormonal urges. Spring migrations can be impressive in numbers, but even without the quantities of individuals, the yearly event is extraordinary. The Small-mouthed Salamander moves mostly at night, so unless you're in for an evening's entertainment you will probably miss it.

ID: dark brown to black on top with grey to grey-yellow speckles; black belly with a few light highlights; often markings on flanks and tail resembling lichens; small head and mouth compared to other mole salamanders.

LENGTH: 10–20 cm (4–8").

DISTRIBUTION: in Canada, only on Pelee Island in Lake Erie.

HABITAT: in ponds and under rocks in abandoned gravel pits.

ACTIVITY PATTERNS: mostly active on warm nights during the brief breeding season; at other times of the year, terrestrial and spends much of its time underground.

REPRODUCTION: courts and breeds in ponds and ditches during spring; female lays 200–700 eggs in several sausage-shaped masses that are attached to vegetation or rocks; eggs hatch in 9–10 days and tadpoles transform into their adult form about 3 months later; sexual maturity is reached after 2 years.

FOOD: insects and other invertebrates.

SIMILAR SPECIES: *Blue-spotted Salamander* (p. 122): pronounced blue flecks on dark body. *Spotted Salamander* (p. 126): yellow spots on dark body.

FRENCH NAME: Salamandre à nez court

DID YOU KNOW? The Small-mouthed Salamander often shares breeding ponds with Spotted Salamanders. The Small-mouthed Salamander is listed as Endangered by COSEWIC.

Tiger Salamander

Ambystoma tigrinum

Gray subspecies

Tiger Salamanders' popularity may be due to their permanent grins or their admirable bulldog disposition. Tiger Salamanders live an inconspicuous life from most humans, seasonally wandering from underground burrow to breeding pond and back under the cover of darkness or rain. • Like many mole salamanders, the burrowing tendency of Tiger Salamanders is a necessity whenever not in water. These salamanders occur on the Great Plains of Canada, and the dry prairie can suck the moisture from their skin very quickly. When travelling to and from their breeding sites, loose soil provided by burrowing animals is likely more important than food. Unlike most herpetologists who flip rocks and logs in the search for salamanders, prairie herpetologists dig through pocket gopher mounds in search of their only salamander; they find few. • Pocket gopher mounds are vital pit stops in migrations and critical overwintering sites, but they aren't the only places Tiger Salamanders find themselves tunnelling into. On occasion, they will find relief from the summer heat and winter cold in the earthen basement of a farmhouse or other underground storage area. When detected, salamanders rarely

SIMILAR SPECIES

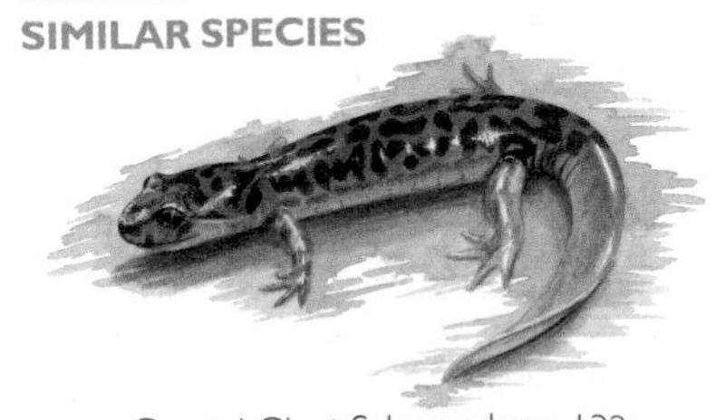

Coastal Giant Salamander, p. 132

suffer the same persecution as similarly motivated mice or snakes. Even the tigers of the salamanders seem to have inherent charm.

ID: large, stout body with stocky legs; tail equals half of total body length; small eyes; rounded snout. *Blotched* subspecies: British Columbia, Alberta and Saskatchewan; light green to brown body with dark stripes or spots. *Gray* subspecies: Saskatchewan and Manitoba; light grey or yellowish body with small, rounded, black spots. *Eastern* subspecies: southeastern Manitoba; dark body with light spots or blotches.

LENGTH: 15–25 cm (6–10"); can grow to 30 cm (12"). *Male:* longer tail and hind legs.

DISTRIBUTION: Okanagan Valley of British Columbia; Alberta, Saskatchewan and Manitoba prairie. *Selected Sites:* Cypress Hills Interprovincial Park (AB and SK), Riding Mountain NP (MB), Grasslands NP (SK), Okanagan Valley (BC).

HABITAT: prairie, aspen parkland and boreal forest; litter and burrows near water.

ACTIVITY PATTERNS: primarily nocturnal during their brief period of annual activity; bury themselves underground from September to April.

REPRODUCTION: spring rains start the migration to breeding ponds; breeding occurs in permanent or semi-permanent ponds or lakes; male courts female and deposits a sperm package that looks like a golf ball on a tee; female picks this up with her cloaca; up to 5000 eggs are laid under stones or on aquatic vegetation, individually or in small groups; eggs hatch within 1 month; tadpoles transform approximately 1 month later, depending on water temperature and food availability.

FOOD: *Adult:* earthworms and other slow-moving invertebrates. *Tadpole:* aquatic invertebrates, larval amphibians, fishes and even other salamanders.

SIMILAR SPECIES: *Coastal Giant Salamander* (p. 132): not found within the Tiger Salamander's range.

FRENCH NAME: Salamandre tigrée

DID YOU KNOW? In certain lakes and wetlands, Tiger Salamanders may live their entire lives without transforming into terrestrial adults. The neotenous Tiger Salamanders attain sexual maturity, but they retain the feathery gills typically found only on the tadpoles. Neoteny usually occurs when the aquatic environment is safer than the terrestrial environment. Tiger Salamanders are listed as Endangered in British Columbia by COSEWIC.

Coastal Giant Salamander

Dicamptodon tenebrosus

Two things may occur in the search for Coastal Giant Salamanders. The least likely is actually finding these large amphibians; the most likely is slipping on rocks and falling into toe-numbing water. Coastal Giant Salamanders spend much of their lives in and around the fast-flowing streams that rush down through the coastal mountains that frame British Columbia's Fraser Valley. • Immature salamanders are most often encountered by gently flipping flat rocks in streams, but terrestrial adults have been encountered precious few times north of the 49th parallel. In Canada, many mature adults retain their external gills and do not venture onto land. • Coastal Giant Salamanders are quite formidable, both as predators and prey. As the largest terrestrial salamanders in Canada, they can prey upon species not normally eaten by amphibians. Typical meals are of the invertebrate variety, but on occasion these behemoths will dine on mice and voles. Coastal Giants possess a strong killing bite that will become apparent if you place a finger or toe within their strike zone. • Coastal Giant Salamanders have experienced a sizeable range reduction in Canada, as

SIMILAR SPECIES

Tiger Salamander, p. 130

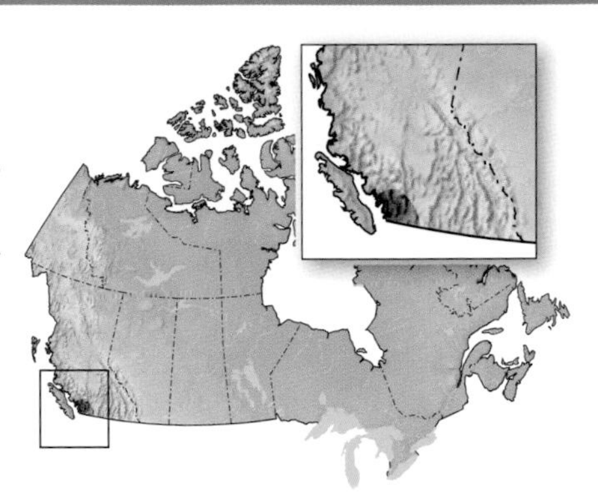

they live in an area that received little protection historically. Streams without fish were traditionally viewed as having little value; thus many forestry activities were permitted, which directly affected water quality and salamander populations.

ID: grey or brown, smooth skin with black marbling; blunt snout; light-coloured underside; older salamanders have faded patterns and are darker overall.

LENGTH: 17–30 cm (6½–12"); males tend to have longer tails.

DISTRIBUTION: area only around Chilliwack, British Columbia.

HABITAT: unlogged or well-established second growth Douglas-fir and broadleaf maple forests. *Terrestrial adult:* underneath mosses, logs and rocks near streams and lakes. *Tadpole:* beneath flat rocks in streams; fast-flowing streams have many rocks, logs and waterfalls that prevent predacious fish from occupying the same habitat.

ACTIVITY PATTERNS: adults are active at night or on rainy days; tadpoles and adults may be most active from April to September.

REPRODUCTION: following a poorly known courtship activity, females lay 75–100 large eggs individually under rocks or logs in a stream; eggs each have individual stalks that attach the egg to the substrate, preventing them from being washed downstream by the current; females guard the eggs until hatching in fall; tadpoles may not transform for 2–3 years and may remain neotenous as adults.

FOOD: *Terrestrial adult:* worms, slugs, other invertebrates, frogs, other forest floor salamanders, small gartersnakes and occasionally small mammals. *Tadpole:* aquatic insects, particularly stonefly and mayfly larvae.

SIMILAR SPECIES: *Tiger Salamander* (p. 130): not found within the Coastal Giant Salamander's range.

FRENCH NAME: Grande salamandre de côtière

DID YOU KNOW? When frightened, Coastal Giant Salamanders "bark." They are the only salamanders in Canada with a true voice. In defence, they will also secrete a distasteful milky solution from their tail. Coastal Giant Salamanders are listed as Threatened by COSEWIC.

Wandering Salamander

Aneides vagrans

Refreshingly agile, strong and quick, this West Coast specialist defies salamander convention. The streamlined Wandering Salamander puts its little legs into high gear when discovered, running or jumping away from a threat. The Wandering Salamander further rebels against salamander conformity by climbing trees up to 40 metres (130 feet)! • The Vancouver Island populations of the Wandering Salamander tell a story that spans thousands of years. Field observations indicate that this species is conspicuously absent from mainland British Columbia and Washington State. In fact, the closest known population to the Canadian Wandering Salamander is found along the coast in northern California. Climate has changed over the millennia, and at times the Wandering Salamander likely had a more continuous range. With lower ocean levels, a land bridge spanned the Olympic Peninsula and Vancouver Island, which allowed species such as the Wandering Salamander to expand north. As the climate changed, the land bridge was flooded and the island population became discontinuous. Still, the huge gap from Vancouver Island to California is hard to explain.

SIMILAR SPECIES

Ensatina, p. 140

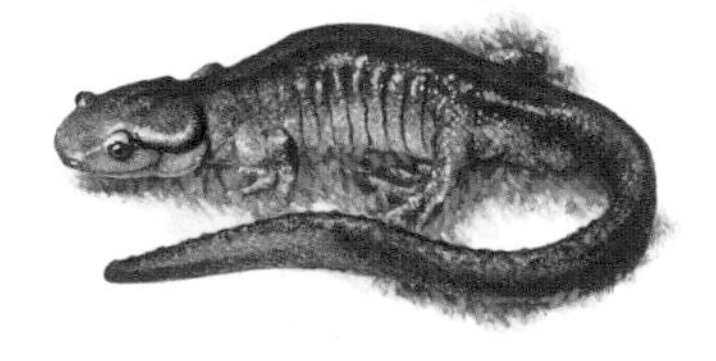

Northwestern Salamander, p. 118

ID: dark brown to grey with brassy or coppery flecks; flecks tend to be concentrated on tail, snout and legs, but can be more widespread to give salamander a lighter appearance; slender and long-legged with squarish-tipped toes; belly is lighter than back.

LENGTH: 7–13 cm (3–5").

DISTRIBUTION: Vancouver Island and the Gulf Islands. *Selected Sites:* Pacific Rim NP, Ripple Rock trail (both in BC).

HABITAT: coastal forests, particularly Douglas-fir; principally in areas with large, rotting logs and talus; most common in mature forests, but may also be found under debris in regenerating clearcuts, burned areas, rocky areas or other clearings.

ACTIVITY PATTERNS: most active at night or during rains from April to October; not commonly seen in the open; entirely terrestrial.

REPRODUCTION: breeds in late spring or early summer, although the female's follicles enlarge in fall in preparation; female lays 8–25 eggs on the roof of a cavity in a rotting log; eggs are individually attached by a jelly-like stalk; female guards the eggs until they hatch; no free-living larval stage; embryos have gills until just before they hatch but leave the egg as miniature adults.

FOOD: all forms of terrestrial invertebrates—black ants, mites, beetles, flies and sowbugs.

SIMILAR SPECIES: *Ensatina* (p. 140): tail constricted at base; black spots on belly. *Northwestern Salamander* (p. 118): larger with obvious glands on back of neck.

FRENCH NAME: Salamandre errant

DID YOU KNOW? The Wandering Salamander was once classified as a Clouded Salamander but was recognized as a distinct species in 1998.

Northern Dusky Salamander

Desmognathus fuscus

Northern Dusky Salamanders sit out late at night, perched in the spray of low-flow waterfalls. As they bathe in the mist and stare distantly into the dark, they give the impression of ruling over their chosen domain. Should the beam of a flashlight disturb this sight, the threatened salamanders will turn tail and head for cover. They retreat by squirming under flat rocks, into pools and behind moss curtains to escape probing fingers. Northern Dusky Salamanders evade capture with greater effectiveness than just about any other salamander because they have the ability to find hiding places where none seem to be. • The effectiveness with which Northern Dusky Salamanders elude threats may be the consequence of their home-bound habits. These salamanders are not world travellers—their territories seem to be little larger than a kitchen table, although they may wander a little farther up and down streams in the course of their lives. They use

SIMILAR SPECIES

Northern Two-lined Salamander, p. 142

Allegheny Mountain Dusky Salamander, p. 138

their keeled, fin-like tails to effectively wiggle fish-like when in a rush, which is further evidence that they prefer water. On land Northern Dusky Salamanders concentrate their activities to the first few hours following nightfall. On humid nights they may remain active until the early hours of morning, but after a few hours on land, they crawl back under the sanctuary of their streamside rock.

ID: variably coloured and patterned; tan to dark brown body with mottles or no markings; light line runs from eye to jaw; tail is keeled and triangular; light-coloured dorsal stripe bordered by darker pigments; old individuals tend to be darker. *Male:* larger.

LENGTH: 5–14 cm (2–5½").

DISTRIBUTION: south of the St. Lawrence River in Quebec into southern New Brunswick; also found at one site near Niagara Falls. *Selected Sites:* Fundy NP (NB), Niagara Gorge (ON).

HABITAT: spring-fed creeks that flow through forests; most frequently discovered under large, flat rocks that lie at the waterline.

ACTIVITY PATTERNS: active when stream temperatures rise above 7° C (45° F); active from May to October; in summer, active during wet, humid nights; in winter, move into underground burrows.

REPRODUCTION: court and breed in spring; females will lay a grape-like cluster of up to 30 eggs near water under a rock or crevice; females guard the eggs until they hatch in 2–3 months; aquatic tadpoles transform into adults the following year; sexual maturity is reached in 3–4 additional years.

FOOD: aquatic and terrestrial insects, centipedes, millipedes and spiders; the larger the salamanders become, the larger the food they eat.

SIMILAR SPECIES: *Northern Two-lined Salamander* (p. 142): 2 stripes down back; lacks light, diagonal eye line. *Allegheny Mountain Dusky Salamander* (p. 138): rounded tail without a keel.

FRENCH NAME: Salamandre sombre du nord

DID YOU KNOW? Northern Dusky Salamanders do not have obvious toxic defences, so they rely on evasion as a defence. The Spring Salamander is a major predator of them throughout much of their range, but these animals occur together in only a few streams here in Canada.

Allegheny Mountain Dusky Salamander

Desmognathus ochrophaeus

Covey Hill in southern Quebec was discovered to hold the only Canadian population of this stream-loving salamander in 1988. That can be considered a surprising discovery given the fact that herpetologists have been scouring the Canadian countryside for almost 200 years. But then again, it's not that much of a surprise, because the Allegheny Mountain Dusky Salamander is a one of the hardest salamanders to find even in the heart of its range in the Appalachians. Therefore, herpetologists were delighted when a second small population was discovered in 2004 in the Niagara Gorge in Ontario. • News of these discoveries spread like wildfire through Canadian herpetological circles. A fine effort has been undertaken in the interim to grasp the significance of the two populations. The Quebec population is small and limited to only 10 sites in a 50-square-kilometre (19-square-mile) region of our vast country. Along the best stretches of river habitat, Allegheny Mountain Dusky Salamanders were found under rocks and logs every metre or so. • Although this salamander is not widespread in Canada, it is quite keen at defending its turf. This little salamander objects to Eastern Red-backed intruders, such that most of the prime streamside shelter is claimed and defended by these marginally Canadian amphibians.

SIMILAR SPECIES

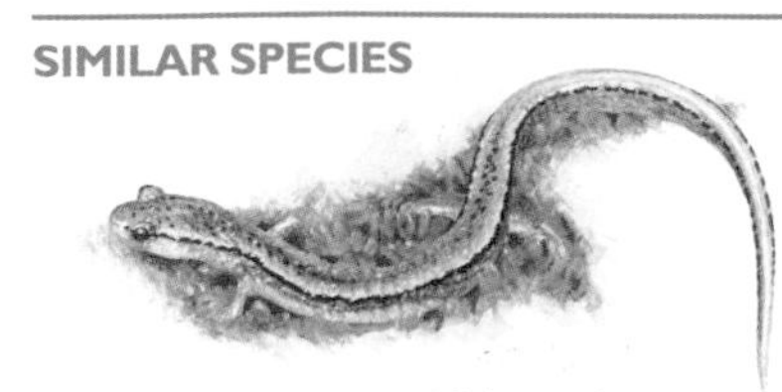

Northern Two-lined Salamander, p. 142

Northern Dusky Salamander, p. 136

ID: wide, brown-yellow or red band bordered by a darker line running down back; occasionally has a row of dark, V-shaped marks running along dorsal stripe; light line runs from eye to jaw; tail is rounded; belly is dark brown to black; lower sides are mottled.

LENGTH: 7–11 cm (3–4 1/2").

DISTRIBUTION: only 2 known locations: Covey Hill in southern Quebec and the Niagara Gorge in Ontario.

HABITAT: along spring-fed creeks; may move into forested areas during the height of summer, but returns to the seepages and springs to hibernate.

ACTIVITY PATTERNS: tends to be active mostly at night, but may move about during the day as well; may be active within the water during winter; probably hibernates from October to April.

REPRODUCTION: mates in pring; approximately 12 eggs are laid in clusters on moss or rotting logs near streams; female guards eggs; tadpoles develop in the water for 3 weeks to 8 months; transform into adult form the next summer; mature in 3–4 years.

FOOD: aquatic and terrestrial insects and invertebrates found among the rocks along streams; also snails, earthworms, mites, caterpillars and flies.

SIMILAR SPECIES: *Northern Two-lined Salamander* (p. 142): yellow band with dark borders down back; lacks light line from eye to jaw. *Northern Dusky Salamander* (p. 136): keeled or triangular tail.

FRENCH NAME: Salamandre sombre des montagnes

DID YOU KNOW? When attacked by a gartersnake, the Allegheny Mountain Dusky Salamander fights back by biting at the attacker's mouth. The Allegheny Mountain Dusky Salamander is listed as Threatened by COSEWIC.

Ensatina

Ensatina eschscholtzii

The subspecies of *Ensatina eschscholtzii* that occurs in Canada is *oregonensis*, a spunky, orange-brown salamander with buggy eyes and a pinched tail. It is a fully terrestrial salamander that needs moisture, but it avoids ponds, creeks and pools of every kind in favour of damp leaves and moss. Six other subspecies of this "super species complex" occur farther south down the Pacific coast, and all seven of the subspecies can be found in California. • The mating walk of this salamander begins with the male nudging the female with his head and neck. If the female is keen, she will lay her head on the male's back and straddle his tail with her back legs. In this position they will walk until the male deposits a sperm packet on the ground. With a little coaxing from the male, the female will pick it up with her cloaca and store it until she is ready to lay her eggs. • The Ensatina does not have a tadpole stage; rather, the young spend considerably more time developing in the egg under the supervision of their mother. This investment of the mother's time comes at the expense of quantity, as fewer than a dozen eggs are laid. • If threatened, the Ensatina will strike a stiff-legged pose with its tail straight and swaying. This movement encourages predators to go for the tail, which will snap off and allow the salamander to escape. The tail also secretes a poisonous, white substance,

SIMILAR SPECIES

Wandering Salamander, p. 134

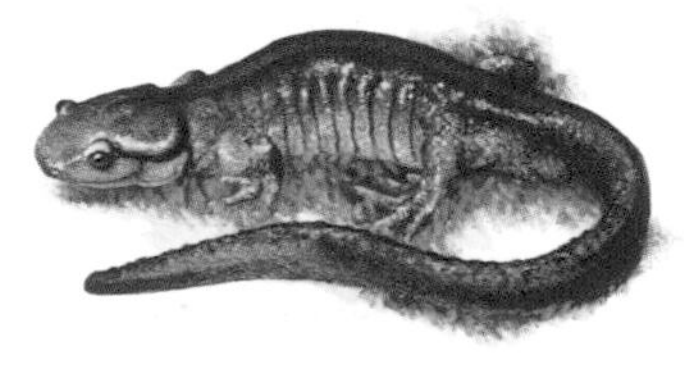

Northwestern Salamander, p. 118

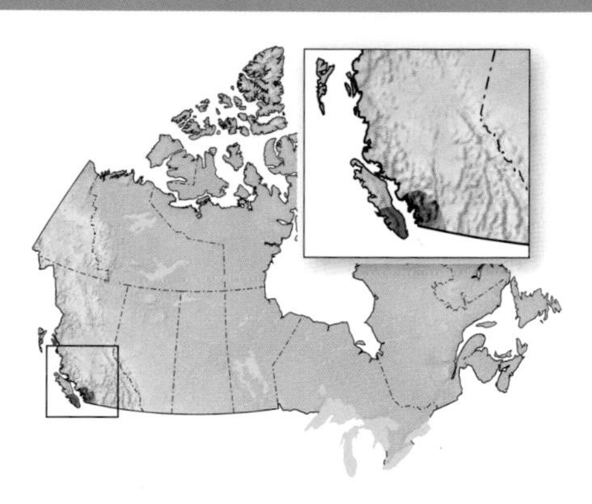

giving the predator something to think about the next time it sees an Ensatina.

ID: skinny, blackish-brown body, sometimes with tiny, white speckles; bulging eyes; 5 toes on hind feet; tail is constricted at base, but if tail has already been lost, replacement tail may not be constricted. *Male:* tail longer than body. *Female:* tail shorter than body.

LENGTH: 7–15 cm (3–6").

DISTRIBUTION: southwestern mainland British Columbia and southern Vancouver Island. *Selected Sites:* Fraser River Valley, southeastern Vancouver Island (both in BC).

HABITAT: moist woodlands, usually in Douglas-fir and vine maple forests; needs nooks and crannies, loose bark and moist ground litter to hide in; will also spend time underground.

ACTIVITY PATTERNS: emerges after rain; burrows into leaf litter or moss during dry spells; retreats underground during colder months.

REPRODUCTION: in early spring (after the first rains), male deposits sperm packet on ground and female picks it up with her cloaca; 2 weeks later female internally fertilizes eggs and lays them in a moist, covered area; female guards eggs until they hatch in early fall; young are developed salamanders; mature in 2–3 years; may live to 15 years.

FOOD: insects, beetles, crickets, ants and other invertebrates.

SIMILAR SPECIES: *Wandering Salamander* (p. 134): no constriction on tail; white dots on underside. *Northwestern Salamander* (p. 118): larger with obvious glands on back of neck.

FRENCH NAME: Salamandre variable

DID YOU KNOW? Lungless salamanders such as the Ensatina breathe through their skin, so picking them up can be quite harmful to them, especially if you have bug spray or lotion on your hands.

Northern Two-lined Salamander

Eurycea bislineata

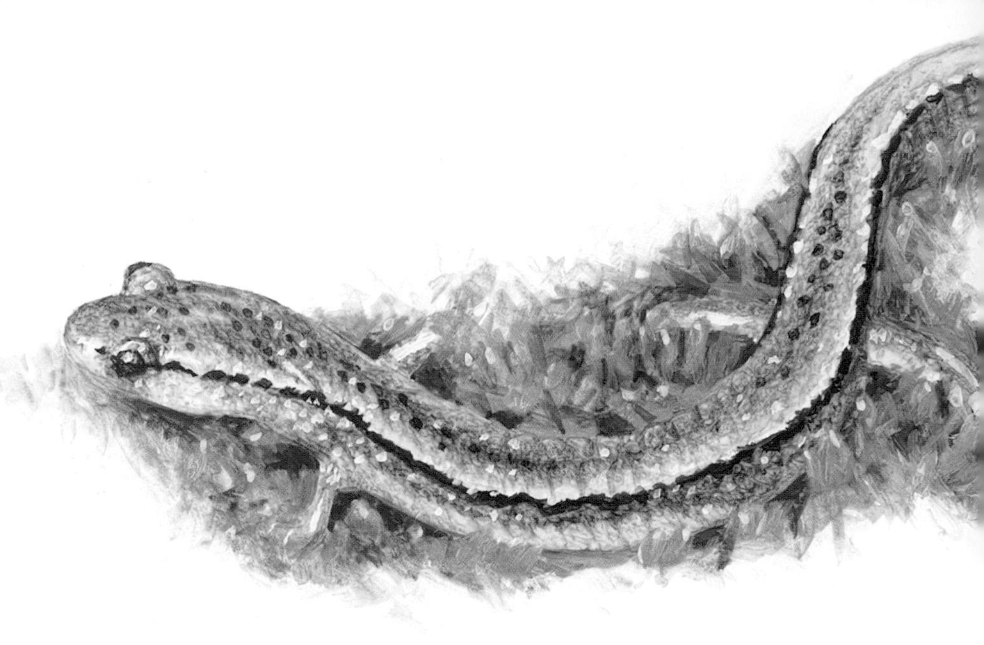

The two lines referred to in the Northern Two-lined Salamander's name are overshadowed by the thick, yellow line running down its back. The two lines are actually the row of dense, black spots along each side of the yellow line. The yellow line continues down the powerful tail, which makes up half of this salamander's length. This long tail aids the Northern Two-lined Salamander in swimming through the cool, fast-flowing streams it calls home. • The male Northern Two-lined Salamander has discovered that the way to a female's heart is through massage. During courtship, the male will lightly scratch the back of the female with his teeth, encouraging her to begin a tail-straddling walk during which he will deposit a spermatophore on the ground, and she will pick it up with her cloaca.

ID: long, thin body; tail makes up half of length; thick, yellow stripe along back of entire body including head; stripe tends to become brighter along tail; stripe may have dark spots within

SIMILAR SPECIES

Northern Dusky Salamander, p. 136

Allegheny Mountain Dusky Salamander, p. 138

it; stripe bordered by a black, broken line on each side; delicate, small legs.

LENGTH: 6–12 cm (2 1/2–5").

DISTRIBUTION: from southern Quebec to New Brunswick; Labrador; southeastern Ontario. *Selected Sites:* Algonquin PP (ON), almost any rocky stream within its range.

HABITAT: cool, moist edges of fast-flowing streams in forested areas; prefers sides of streams with leaf litter, rocks and vegetation that provide hiding places.

ACTIVITY PATTERNS: primarily nocturnal; prefers to stay out of the sun as it can dry out quickly; remains dormant in colder months by staying underground or under thick leaf litter.

REPRODUCTION: breeds in early spring; female lays 15–100 eggs under rocks that may be within the water; female guards eggs until they hatch in 1–2 months; young remain as aquatic tadpoles for 2–3 years; mature the first year after transformation into adults.

FOOD: *Adult:* terrestrial invertebrates found along the shores of fast-flowing streams; ants, mites and spiders. *Tadpole:* aquatic invertebrate larvae.

SIMILAR SPECIES: *Northern Dusky Salamander* (p. 136) and *Allegheny Mountain Dusky Salamander* (p. 138): thicker body; shorter tail; light stripe runs from eye to jaw line.

FRENCH NAME: Salamandre à deux lignes

DID YOU KNOW? The tadpoles of the Northern Two-lined Salamander can sense when a predatory fish is in their area—the tadpoles will then hide under rocks or debris in the water, sometimes not emerging until a couple of days later!

Spring Salamander

Gyrinophilus porphyriticus

Whether you are a fashion model, a tropical bird or a salamander, bright colours have the same purpose—to grab attention. Although some animals (ourselves included) may use a vivid wardrobe to proclaim our reproductive desirability, this is not the aim of Spring Salamanders. For these colour-blind amphibians, the assessment of their suitors is done through dance and displays, leaving the flaming fashion statement for predators' eyes alone. • The red efts of Eastern Newts are the only other red animals in this book, and both they and Spring Salamanders wear the colour with similar purpose. Both have noxious toxins in their skin that predators learn to avoid at first taste. The mimicking teamwork also includes Red Salamanders, a species that lives as far north as Lake Erie's southern shore and overlaps considerably with the other two throughout the eastern United States. It is believed that all three mimic one another to teach their potential threats to leave one and all alone. • Spring Salamanders are rare in Canada. Recent findings have discovered two populations in southern Quebec. Here at the northern fringe of their entire range it is thought that fewer than one thousand Spring Salamanders can boast Canadian citizenship.

SIMILAR SPECIES

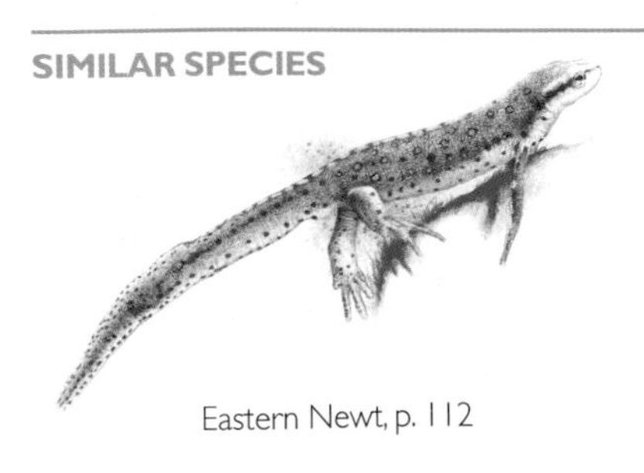

Eastern Newt, p. 112

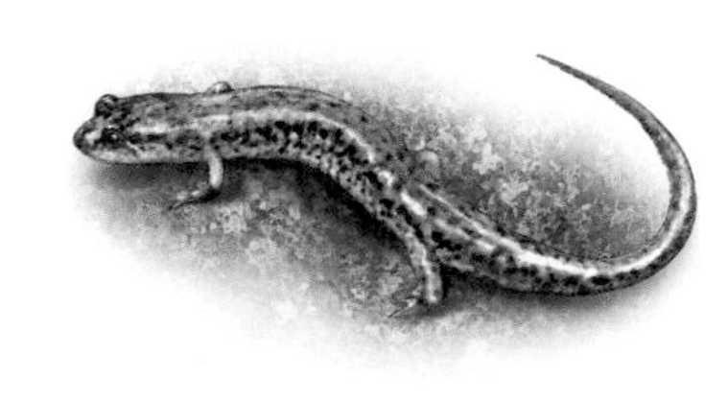

Northern Dusky Salamander, p. 136

ID: large; reddish body; black, spotted or net-like pattern on back; light line reaches from eye to nostril; keeled, fin-like tail; light belly. *Adult:* becomes increasingly dark with age.

LENGTH: 11–21 cm (4–8").

DISTRIBUTION: only in southern Quebec; 2 known Canadian populations are separated by the Richelieu River/Lake Champlain/Hudson River system; may have formerly been found in Ontario in the Niagara Gorge. *Selected Sites:* mountains of southeastern Quebec.

HABITAT: cool, well-oxygenated, clear streams in forested, mountainous regions; during summer, adults are usually found underneath large, flat rocks at waterline; in winter, they are found in springs.

SIMILAR SPECIES

Four-toed Salamander, p. 146

ACTIVITY PATTERNS: nocturnal; spend winter tucked into springs or seepages from October to May.

REPRODUCTION: mate in fall; females lay eggs the following spring or summer; eggs are attached to the underside of a large rock or log in streams; eggs hatch in fall; may take up to 4 years before tadpoles transform into adults.

FOOD: other salamanders; terrestrial and aquatic invertebrates.

SIMILAR SPECIES: no other Canadian salamander is the same colour. *Eastern Newt (eft phase)* (p. 112): much smaller; rounded tail. *Northern Dusky Salamander* (p. 136): light line on face from eye to corner of mouth; much smaller. *Four-toed Salamander* (p. 146): smaller; base of tail constricted; 4 toes on hind feet.

FRENCH NAME: Salamandre pourpre

DID YOU KNOW? Spring Salamanders have a fierce reputation as predators of other, smaller salamanders. Spring Salamanders are listed as Special Concern by COSEWIC.

Four-toed Salamander

Hemidactylium scutatum

Bogs are the sole home of the Four-toed Salamander, but rather than inhabiting our unending selection of northern bogs, this amphibian is found only in boggy habitats that hug the American border. A bog is a fine place for a salamander. The underlying sphagnum moss found there retains water so effectively that the amphibian is always basking in humid, steamroom comfort. The moss also provides a labyrinth of travel opportunities for an animal inclined to remain undetected. Its bog habitat also, in part, is why so little is known about the Four-toed Salamander in Canada. It lives in a place that is unfriendly to humans, and it is a master at hide and seek. This combination, however, does drive certain challenge-loving herpetologists to look for it, resulting in significant new localities being discovered every few years. • The Four-toed Salamander belongs to the family of lungless salamanders, but it does not stay altogether loyal to family ties. It does breathe exclusively through its skin and lay its eggs on land, but unlike some other lungless salamanders that are born fully formed, the Four-toed

SIMILAR SPECIES

Spring Salamander, p. 144

Eastern Red-backed Salamander, p. 150

Salamander hatches out partly formed, similar to the tadpole of a frog or a mole salamander. Once hatched, the tadpole wiggles itself free of the nest and drops into a small pool of water, where it remains until it has finished developing.

ID: 4 toes on hind feet; rusty brown back; grey sides; belly is white with a few black blotches; base of tail is slightly constricted. *Female*: larger than male; proportionally shorter tail.

LENGTH: 5–10 cm (2–4").

DISTRIBUTION: disjunct patches in southern Ontario, Quebec, Nova Scotia and New Brunswick. *Selected Sites:* Fundy NP (NB), Kejimkujik NP (NS).

HABITAT: bogs with sphagnum moss; also nearby hardwood forests along flood plains or streams in woodland areas.

ACTIVITY PATTERNS: emerges from hibernation in April and remains relatively active as late as early November; most wandering occurs at night.

REPRODUCTION: mates in fall; in early spring, up to 30 eggs are laid in a sphagnum moss nest; more than one female may share a nesting site; females guard the eggs that hatch in 1–2 months; newly hatched young move to small pools where they finish their larval development; sexually mature in approximately 2 years.

FOOD: beetles, moths, spiders and mites.

SIMILAR SPECIES: *Spring Salamander* (p. 144): light bar from eye to nostril. *Eastern Red-backed Salamander* (p. 150): 5 toes on hind feet; base of tail not constricted; lacks white belly with black spots.

FRENCH NAME: Salamandre à quatre doigts

DID YOU KNOW? Most salamanders have four toes on the front feet but five on the hind feet. The Four-toed Salamander is one of the few salamanders with only 16 toes.

Coeur d'Alene Salamander

Plethodon idahoensis

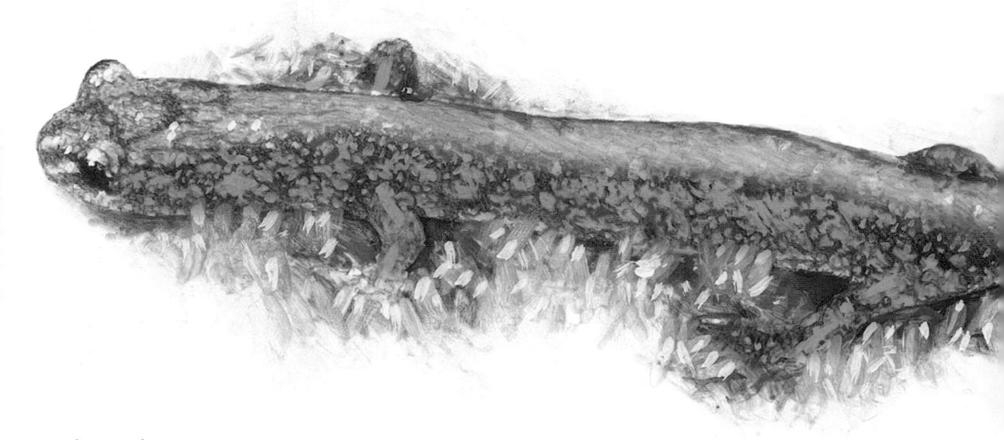

The hills around Creston, British Columbia are the extreme northern range of the Coeur d'Alene Salamander. Its U.S. range is not much more generous—it occurs in only small parts of both Idaho and Montana. Only a select few Canadians have ever seen this terrestrial salamander. It is most active during spring rains or at night, and any search effort can do much damage to its remaining, highly sensitive habitat. Although it was only discovered in Canada in 1981, the Coeur d'Alene Salamander is a declining species in this country. The best thing caring naturalists can do for this animal is to celebrate its citizenship without actually disturbing it. • The Coeur d'Alene Salamander is thought by some experts to be the same species as Van Dyke's Salamander, a rusty-coloured resident of Washington. The two species are believed to have been separated by glaciation. • The Coeur d'Alene Salamander, like

SIMILAR SPECIES

Western Red-backed Salamander, p. 152

Long-toed Salamander, p. 124

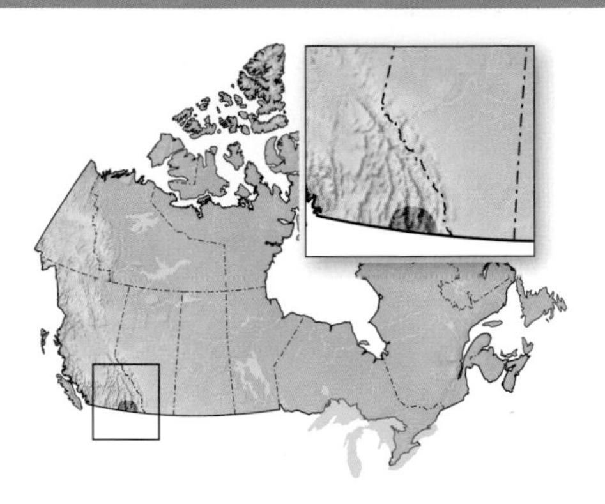

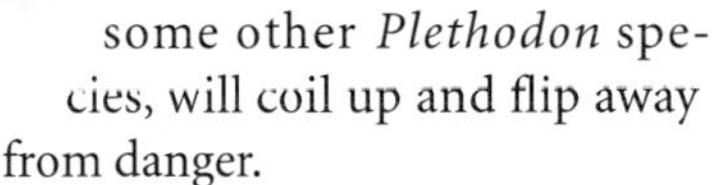

some other *Plethodon* species, will coil up and flip away from danger.

ID: obvious gland on top of head; yellow patch on throat; body colour varies from black, tan, yellow or reddish; wavy tan, yellowish or reddish stripe on back; cream-coloured spots on entire underside.

LENGTH: 6–12 cm (2½–5").

DISTRIBUTION: restricted to the southeastern side of Kootenay Lake and the Columbia River Valley in southern British Columbia.

HABITAT: along streams, on slopes and in moist, cool, conifer forests; likes seeps.

ACTIVITY PATTERNS: active from March or April to October in southern British Columbia; ventures out in warm spring rains but remains under forest debris for much of the remaining season of activity.

REPRODUCTION: little known; like other lungless salamanders, a grape-like cluster of 4–12 eggs is likely laid in a moist place like the inside of a log, hung from the "ceiling" by a jelly-like strand; female guards eggs; hatchlings are miniature (3–5 cm [1–2"]) replicas of adults.

FOOD: little known; presumably invertebrates.

SIMILAR SPECIES: *Western Red-backed Salamander* (p. 152): lacks yellow throat; border of back stripe is distinct; ranges do not overlap. *Long-toed Salamander* (p. 124): lacks both parotid gland on head and yellow throat.

FRENCH NAME: Salamandre de Coeur d'Alene

DID YOU KNOW? *Plethodon* means "many teeth," a tribute to the row of teeth found in all members of this genus. The Coeur d'Alene Salamander is listed as Special Concern by COSEWIC.

Eastern Red-backed Salamander

Plethodon cinereus

If you wandered over to a small woodlot on Prince Edward Island and turned over a rotting piece of wood, more often than not you would meet an Eastern Red-backed Salamander. This particular salamander is the most conspicuous lungless salamander in eastern Canada, and it may be the most frequently encountered salamander in the entire country. It is a small, skinny and dainty animal that seems more worm-like than anything else, but the feeble legs and slender body mask a very resilient animal. Given sufficient observation, the charm of this salamander soon emerges as it walks with robotic regularity and occasionally flexes its body to jump slight distances. • As a result of its wide distribution and abundance, the life history of this species is quite well known. The Eastern Red-backed Salamander becomes active with spring rains and forages at night on the ground and on vegetation. As the

SIMILAR SPECIES

Northern Dusky Salamander, p. 136 | Four-toed Salamander, p. 146 | Western Red-backed Salamander, p. 152

warm weather persists and the land begins to dry, the salamander retreats first into rotting logs or depressions, finally retreating into the ground itself before the surface completely dries. • The Eastern Red-backed Salamander is territorial, and larger individuals tend to occupy habitat that is rich in prey species. These large individuals are also good at defending quality sources of cover, and large males attract more females than do smaller individuals. The salamander marks its territory with feces and secretions from glands on its tail and shoulders.

ID: long and skinny; tiny legs. Two colour forms occur: most common is black or dark grey with a thick, reddish, orange, grey, yellow or pink stripe along entire top of body; less common "Leadback" form lacks stripe and is therefore all black or grey; both forms tend to have black and white blotches on belly.

LENGTH: 6–13 cm (2 1/2–5").

DISTRIBUTION: from the Great Lakes to the Maritimes. *Selected Sites:* Algonquin PP (ON), Pukaskwa NP (ON), La Mauricie NP (QC), Fundy NP (NB), Kouchibouguac NP (NB), Cape Breton Highlands NP (NS), Kejimkujik NP (NS).

HABITAT: mature, deciduous or mixed forests; beneath fallen logs and coarse, woody ground litter.

ACTIVITY PATTERNS: active from mid-spring to mid-fall; occasionally seen moving on warm and rainy spring nights; spends most of its life hidden beneath forest debris; spends winter in underground burrows.

REPRODUCTION: breeds in fall; female lays 3–13 eggs in spring and hangs them in a clump from the ceiling of a damp stump or log; eggs are guarded by the female for 6–8 weeks until hatching; female also gives some early parental care to the hatchlings.

FOOD: many invertebrates, including flies, beetles, ants, spiders and worms.

SIMILAR SPECIES: *Northern Dusky Salamander* (p. 136): keeled tail; light line between eye and jaw. *Four-toed Salamander* (p. 146): 4 toes on hind feet; base of tail slightly constricted. *Western Red-backed Salamander* (p. 152): confined to British Columbia.

FRENCH NAME: Salamandre rayée

DID YOU KNOW? When they hatch, young Eastern Red-backed Salamanders have gills that are absorbed into their bodies. They remain with their mother for several days after hatching.

Western Red-backed Salamander

Plethodon vehiculum

A few minutes from Victoria lies the motherlode of salamanders. In Goldstream Provincial Park, the Western Red-backed Salamander steals centre stage from some of Canada's tallest trees and best salmon runs...but only if you bother to flip a log or two. Because it is found in tremendous concentrations in this particular park, you have an excellent chance of meeting the Western Red-backed Salamander simply by turning over a log. It is estimated that in Goldstream this salamander has the largest vertebrate biomass. • Flipping for salamanders is much more productive in a place like Goldstream than in other places, but wherever you are, finding one under a hiding place is always a surprise because predictions of the presence of salamanders are never as successful as expected. The thrill does not stop at the discovery, though. There is always a half-hearted chase as the salamander makes for the nearest cover. Occasionally it does vanish mysteriously into the damp decay, but more commonly its clean, slick form yields itself to study. The salamander's eyes stare without blinking and the

SIMILAR SPECIES

Eastern Red-backed Salamander, p. 150

Coeur d'Alene Salamander, p. 148

Long-toed Salamander, p. 124

snout steers instinctively for crevasses, whether among leaves or between fingers. • When flipping for salamanders, make sure to flip carefully and return the litter to as much of its original state as possible; this keeps important hiding spots for the Western Red-backed Salamander intact.

ID: long body; dark brown to black base colouring, sometimes lighter; broad, red (sometimes yellow, light green or beige) stripe down back; stripe may be absent in some individuals; belly has small, dark and light flecks.

LENGTH: 7–12 cm (3–5").

DISTRIBUTION: Vancouver Island and the Fraser Valley to Hope, British Columbia. *Selected Sites:* Goldstream PP, Cultus Lake PP, Pacific Rim NP (all in BC).

HABITAT: moist, forested habitats; talus slopes, leaf debris, rotting logs and under rocks.

ACTIVITY PATTERNS: active for much of the year except for cold days in December and January; spends much of its time beneath cover; emerges following rains or at night; may move into the ground during dry summers.

REPRODUCTION: courts and breeds in October or November; grape-like cluster of eggs is laid in spring; female guards the eggs; eggs hatch in fall or the following spring; at peak hatching times, the tiny salamanders can be very numerous; mature in about 2 years.

FOOD: small invertebrates such as mites, isopods, millipedes, snails and earthworms.

SIMILAR SPECIES: *Eastern Red-backed Salamander* (p. 150): not found in British Columbia. *Coeur d'Alene Salamander* (p. 148): yellow stripe and yellow throat; range does not extend far enough west to overlap. *Long-toed Salamander* (p. 124): larger; proportionally larger limbs; long toe second from outside on hind feet.

FRENCH NAME: Salamandre à dos rayé

DID YOU KNOW? The Western Red-backed Salamander may share space with one or more other individuals. Although it does recognize olfactory cues from other salamanders, it does not defend territories as aggressively as the Eastern Red-backed Salamander.

Rocky Mountain Tailed Frog

Ascaphus montanus

In 2000, the species previously known as the "Tailed Frog" was split into two distinct species, the Rocky Mountain Tailed Frog and the Coastal Tailed Frog. The Rocky Mountain Tailed Frog has a more restricted range in Canada than its close cousin. As a result, it is more vulnerable to habitat disturbance. It has two distinct populations in British Columbia, both in the Kootenays in the southeastern corner of the

SIMILAR SPECIES

Coastal Tailed Frog, p. 156

Pacific Treefrog, p. 196

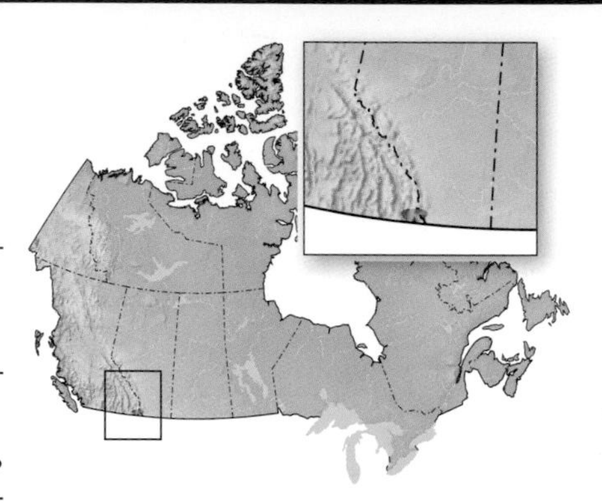

province. One inhabits small streams within a five-kilometre (three-mile) radius, and the other occupies small streams within only a three-kilometre (two-mile) radius. With individuals reproducing only after they reach seven or eight years of age, and with females reproducing only every second year once they are mature, these frogs have a low reproductive output, which makes them even more vulnerable to population decimation. • Tailed frogs produce pea-sized eggs, which are considered the largest among North American frogs. These endearing frogs also carry the designation of being the most primitive frogs in the world!

ID: olive green, tan or dark brown body; distinctive, dark speckling on backs of pale individuals; large head; pale copper-coloured triangle between eye and snout; vertical pupils; pink belly; no external eardrum. *Male:* short but prominent tail.

LENGTH: 3–5 cm (1–2").

DISTRIBUTION: limited to the Kootenays in southeastern British Columbia.

HABITAT: cold, fast-flowing mountain streams in forested areas.

ACTIVITY PATTERNS: adults active at night along the borders of streams; in the coldest weather, frogs and tadpoles have reduced activity and remain hidden under rocks in the streams.

REPRODUCTION: courtship involves no vocalizing, only displaying; mating occurs in fall, with the female retaining the sperm until after spring runoff, when she lays 40–80 eggs in small clusters on the underside of rocks in a stream; after 3–6 weeks hatchlings emerge and overwinter at the hatching site; tadpoles remain in the streams for 3–5 years; sexual maturity is reached at 7–8 years old.

FOOD: *Adult:* forages within 12 m (40') of the water for spiders, flies, caddisflies, beetles and other invertebrates. *Tadpole:* scrapes algae from rocks.

CALL: does not vocalize.

SIMILAR SPECIES: *Coastal Tailed Frog* (p. 156): found only in coastal British Columbia. *Pacific Treefrog* (p. 196): slimmer; round pupils; dark line from nose to shoulder.

FRENCH NAME: Grenouille-à-queue des Rocheuses

DID YOU KNOW? The tongue of this frog doesn't extend out of its mouth the way the tongues of many other frog species do. The Rocky Mountain Tailed Frog must pounce on its prey and capture its food with a quick snap of its mouth. The Rocky Mountain Tailed Frog is listed as Endangered by COSEWIC.

Coastal Tailed Frog

Ascaphus truei

Tailed frogs are unique among Canadian frogs. Coastal Tailed Frogs are well named because the male frogs do, indeed, appear to have tails. The cloaca sticks out (rather than in) on males so they can release their sperm within the female; thus this frog has internal fertilization. Coastal Tailed Frog eyes have catlike, vertical pupils, and they lack the external tympanum (ear membrane) so obvious in other Canadian

SIMILAR SPECIES

Rocky Mountain Tailed Frog, p. 154

Pacific Treefrog, p. 196

frogs. That Coastal Tailed Frogs have poor hearing is not too much of an adaptive disadvantage; they do not vocalize either. Given their continuously noisy environment, vocal communication appears to be one trait best forsaken. • Cool, fast-flowing streams are an unusual habitat for amphibians, but these frogs have chosen to call them home. As tadpoles, Coastal Tailed Frogs have disks around their mouths. This suction structure is used to adhere to rocks so that the tadpoles do not get swept away in the current, and it is also used to scrape off the algae they feed on. • Coastal Tailed Frogs have never had an extensive Canadian range, but they have lost ground in areas of heavy logging. Activities that have disturbed streambeds and increased the silt in waters have caused declines in Coastal Tailed Frog numbers. Current regulations designed to minimize the impact on stream water quality for salmon and other fishes may relieve some of stresses on these stream-dwelling amphibians.

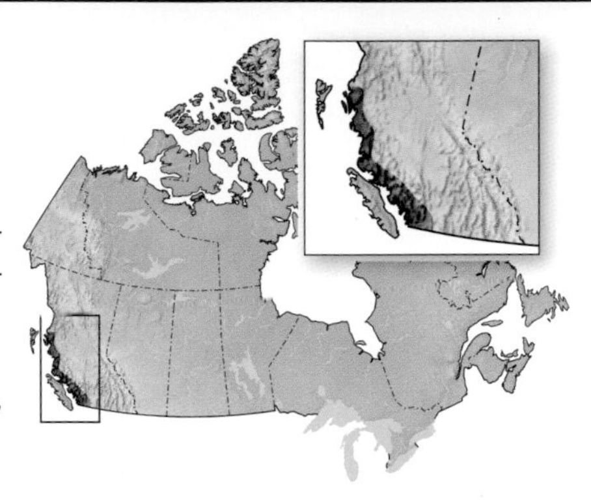

ID: olive green, grey, tan, dark brown or black body; light triangle on head; belly is pink with white spots; vertical pupils; no external eardrum. *Male:* has a short but prominent "tail."

LENGTH: 3–5 cm (1–2").

DISTRIBUTION: coastal British Columbia. *Selected Sites:* Cultus Lake PP, Chilliwack Lake PP (both in BC).

HABITAT: cold, fast-flowing mountain streams in forested areas.

ACTIVITY PATTERNS: adults are active mostly at night; during the coldest weather, frogs and tadpoles have reduced activity and remain hidden under rocks in streams.

REPRODUCTION: courtship involves no vocalizing, only displaying; mating occurs in early fall; female lays 35–100 eggs in small masses on the underside of rocks in a stream the next summer; eggs hatch in 6 weeks; tadpoles remain in the streams for 2–4 years and reach sexual maturity at 8–9 years old.

FOOD: *Adult:* forages in upland areas for invertebrates in the evening. *Tadpole:* scrapes algae from rocks.

CALL: does not vocalize.

SIMILAR SPECIES: *Rocky Mountain Tailed Frog* (p. 154): found only in southeastern British Columbia. *Pacific Treefrog* (p. 196): slimmer; round pupils; dark line from nose to shoulder.

FRENCH NAME: Grenouille-à-queue côtière

DID YOU KNOW? Coastal Tailed Frogs are the longest-lived frogs in Canada; some individuals live up to 20 years! Coastal Tailed Frogs are listed as Special Concern by COSEWIC.

Plains Spadefoot

Spea bombifrons

The first warm rains of spring turn silent prairie nights into short-lived Plains Spadefoot concertos. Shallow, muddy ponds form in slight prairie depressions, flooding drought-tolerant plants and sending a signal down through the parched soil to awaken burrowed Plains Spadefoots. With the stimulation of moisture, the toads scramble to the surface and begin a reproductive race against the prairie sun. Breeding time is what the lives of adult Plains Spadefoots are all about. With the inevitable recession of shallow waters, the toads briefly refuel fat reserves after breeding and mine themselves back into the ground to await the next spring rains. • Plains Spadefoot tadpoles race against the drought clock as well. Confined to an evaporating sanctuary,

SIMILAR SPECIES

Great Plains Toad, p. 166

Canadian Toad, p. 168

Great Basin Spadefoot, p. 160

these tadpoles have a feature that often enables them to beat the shrinking waters. The duration of development and transformation often depends on the availability of water. In less water, toads transform into adults faster but are smaller. In more water, the tadpoles grow slower and transform into larger adults potentially better equipped for survival. Regardless of the time of year of their transformation, it always tends to occur at the last possible moment. Although many tadpoles simply lose the race against the drought, enough young toads survive to make use of their tell-tale spades (see ID) and burrow backward into the soil to await next year's rains.

ID: fat, soft body; light green and brown skin; tiny brown or orange bumps; rounded hump between eyes; vertical pupils; "spade" is a horny, dark knob on soles of hind feet; occasionally light stripes on back.

LENGTH: 4–6 cm ($1^1/_2$–$2^1/_2$").

DISTRIBUTION: the short-grass prairie in southern Alberta and southwestern Saskatchewan; scattered locations in southeastern Saskatchewan and southwestern Manitoba. *Selected Sites:* Dinosaur PP (AB), Writing-on-Stone PP (AB), Grasslands NP (SK).

HABITAT: native grasslands and pastures with loose soils.

ACTIVITY PATTERNS: inactive for much of the year; adults migrate to breeding ponds when heavy rains arrive in late May or June; much of their calling/breeding activity occurs at night.

REPRODUCTION: breeding occurs in shallow prairie ponds after the spring rains; males arrive at breeding sites and call; females lay 10–250 eggs; eggs hatch within 48 hours; tadpoles transform within 3–6 weeks; toads may not breed in very dry years.

FOOD: little is known of their food habits; may opportunistically feed on invertebrates.

CALL: courtship call is a series of short, nasal *waaah, waaah* sounds repeated in 1-second intervals.

SIMILAR SPECIES: *Great Plains Toad* (p. 166): does not have vertical pupils; dark splotches all over body. *Canadian Toad* (p. 168): does not have vertical pupils; dark splotches all over body with larger "warts." *Great Basin Spadefoot* (p. 160): not found outside British Columbia.

FRENCH NAME: Crapaud des plaines

DID YOU KNOW? Some larger Plains Spadefoot tadpoles may resort to cannibalism in order to grow rapidly in stressful situations.

Great Basin Spadefoot

Spea intermontana

Away from the symmetrical orchards and roadside fruit stands that dominate much of the Okanagan Valley lies the parched pocket of drylands that is home for the Great Basin Spadefoot. Few of the millions of tourists that visit the Okanagan Valley for its wine tours and lakes consider the extraordinary ecological significance of this region. This valley is a narrow finger of the Great Basin, a dryland sister to the Great Plains of North America. Only within the steep walls of this valley do many types of wildlife cross into British Columbia, none better named and represented than the Great Basin Spadefoot. • When newly hatched, the mouth parts of the tadpole have yet to develop into the hinged jaws of an adult. Early mouthparts are located on the underside of the head and look

SIMILAR SPECIES

Western Toad, p. 164

Plains Spadefoot, p. 158

much like those of a sucker fish. At this stage the tadpole is a voracious vegetarian and grazes plants and algae with an insatiable appetite. • As the tadpole continues to develop, the mouth gradually begins to appear more adult-like. Throughout this transformation, the tadpole forsakes vegetarianism for carnivorous tendencies. The transformation of tadpole into toad occurs at very different rates. As some tadpoles near the stage of transformation, they often turn on their siblings. Nearly transformed toadlet-tadpoles often consume the less developed individuals for the needed energy to push through to the terrestrial phase. Cannibalism is one of the many processes that has enabled the Great Basin Spadefoot in particular and amphibians in general to become one of the longest-living lines of animals the world has ever seen.

ID: smooth, grey to olive skin with many small, brown or orange tubercles; vertical pupils; glandular hump between eyes; black, keratinous "spade" on soles of hind feet.

LENGTH: 4–6 cm ($1^1/_2$–$2^1/_2$").

DISTRIBUTION: the Okanagan and Thompson/Nicola valleys of British Columbia. *Selected Sites:* White Lake Grasslands Protected Area, South Okanagan Wildlife Management Area, Lac du Bois Grasslands Protected Area (all in BC).

HABITAT: dry valleys with loose soil.

ACTIVITY PATTERNS: burrows into the ground for much of the year; becomes active and digs itself to the surface in early summer to migrate to breeding ponds; primarily active at night, especially during rainy conditions.

REPRODUCTION: breeds in temporary ponds and lakes; male vocalizes at breeding sites; female lays 300–800 black eggs in loose, cylindrical masses of about 40 eggs on vegetation or the bottom of the pond; eggs hatch within a week; tadpoles transform very quickly.

FOOD: *Adult:* invertebrates. *Tadpole:* algae and plant matter.

CALL: a series of short *gwaaaa* sounds; male typically calls in a chorus with other males.

SIMILAR SPECIES: *Western Toad* (p. 164): no "spade" on hind feet; horizontal pupils; light stripe down back. *Plains Spadefoot* (p. 158): not found in British Columbia.

FRENCH NAME: Crapaud du Grand Bassin

DID YOU KNOW? The Great Basin Spadefoot is sometimes brought to the surface by cars or cattle stamping near its burrow. It is speculated that to the buried spadefoot, these sounds mimic the sound of falling rain. The Great Basin Spadefoot is listed as Threatened by COSEWIC.

American Toad
Bufo americanus

Throughout central and eastern Canada, a chorus of amphibian voices resonating from otherwise tranquil ponds is one of nature's most delightful spring traditions. Among the many voices that cry out for female attention is that of the American Toad, whose loud, prolonged trill is easily detected. Once you are able to recognize its voice you will find it nearly impossible to venture outdoors in spring without encountering the American Toad. • Many adults carry with them fond memories of catching toads, frogs and salamanders as children. Although the American Toad is no speedster, its erratic hops into water or shrubs can thwart even the most talented amphibian-catching children. When a

SIMILAR SPECIES

Canadian Toad, p. 168   Fowler's Toad, p. 170

DID YOU KNOW? Both tadpoles and adults have poison glands in the skin. These glands produce toxins to make them less palatable to predators. A taste of an American Toad will make most predators foam at the mouth. However, the Eastern Hog-nosed Snake feeds exclusively on toads and clearly finds them delicious.

toad is cornered or held in hand, its permanent frown, muscular struggles and ritualistic urination do little to tarnish its mystique. • Given the right surroundings, the American Toad is one of the most likely amphibians to be found right in your own backyard. People living in rural and naturally vegetated areas can entice toads into their yards by providing small ponds and hiding spots.

ID: large and stocky with brown, reddish or olive "warty" skin; blunt face; short legs; little webbing between toes of hind feet; whitish line down centre of back; long parotid glands; cranial crests like straight "eyebrows."

LENGTH: 6–11 cm (2½–4½").

DISTRIBUTION: from central Manitoba to Labrador and the Maritimes. *Selected Sites:* Algonquin PP (ON), La Mauricie NP (QC), Kejimkujik NP (NS).

HABITAT: lawns, gardens, fields and forested woodlots.

ACTIVITY PATTERNS: primarily active during twilight and at night; active from early spring to mid-fall.

REPRODUCTION: breeding and courtship occur between late March and early June depending on the latitude; breeding can occur in just about any water that does not have much of a current; up to 20,000 eggs are laid in twin strands and wound around sticks and aquatic vegetation; tadpoles hatch in a few days to a few weeks and transform into tiny terrestrial toadlets in 50–65 days.

FOOD: *Adult:* invertebrates. *Tadpole:* plants and algae.

CALL: each male sings on a different pitch depending on his size; trill is 3–30 seconds long; sometimes a small, single note will precede the singing.

SIMILAR SPECIES: *Canadian Toad* (p. 168): cranial crests come to a "V" between eyes; ranges overlap only in Manitoba. *Fowler's Toad* (p. 170): parotid glands touch cranial crests; found only on Lake Erie shoreline in southern Ontario. American Toads hybridize with both of these species.

FRENCH NAME: Crapaud d'Amerique

Western Toad

Bufo boreas

Western Toads are chunky, soft-spoken inhabitants of western forests. Their easily imitated soft whistles chime from wetlands and streams on early-spring nights. Once day length and breeding duties have naturally silenced their nightly songs, adults move into the forests, where they find a bounty of newly emerged invertebrate life.

• There is great concern over the status of Western Toads throughout much of their distribution in the United States. Severe die-offs have been reported throughout Colorado and Oregon. Fungus infections and increased UV

SIMILAR SPECIES

Canadian Toad, p. 168

Great Basin Spadefoot, p. 160

radiation have been implicated in these declines, but the causes are still uncertain. Fortunately, in Canada there has been little evidence to suggest that our toad populations are in jeopardy. An increase in research has been conducted in response to concerns, and this work seems to indicate that this toad extends farther north and east than previously thought. Some people have speculated that this expansion may be the result of climate change, habitat change or an adaptive advantage over the Canadian Toad.

ID: brown, red or green body; darker "warts"; single, light stripe down middle of back; no cranial or post-orbital crest between or behind eyes; kidney-shaped parotid glands; pale belly with dark marks; gold-flecked eyes with horizontal pupils.

LENGTH: 6–12 cm (2½–5"); females generally larger.

DISTRIBUTION: throughout British Columbia to the southeast corner of the Yukon and east to the Athabasca River in Alberta. *Selected Sites:* Kananaskis Country (AB), Manning PP (BC), Goldstream PP (BC).

HABITAT: coniferous, or less frequently deciduous, forests near beaver ponds, ditches, streams and lakes; breeding takes place in shallow ponds, ditches or lakes with clean water.

ACTIVITY PATTERNS: active from early spring to mid-fall; courtship and breeding activity is chiefly nocturnal.

REPRODUCTION: depending on latitude, courtship and breeding will occur sometime between April and June; males vocalize at night; males mount the females in amplexus while the females lay strings of up to 16,500 eggs; eggs hatch after 3–12 days; tadpoles begin to transform in 6–8 weeks.

FOOD: *Adult:* invertebrates, including insects, spiders and worms. *Tadpole:* plants and algae.

CALL: courtship call is lacking, or in some Alberta populations may recur as a quiet series of soft whistles.

SIMILAR SPECIES: *Canadian Toad* (p. 168): ranges overlap only in Alberta; cranial crests form ridge between eyes. *Great Basin Spadefoot* (p. 160): south-central British Columbia only; vertical pupils; no line down back.

FRENCH NAME: Crapaud de l'ouest

DID YOU KNOW? Western Toads are not hoppers—they prefer to walk. Western Toads seem to be doing well in Canada but have been listed as Special Concern by COSEWIC because of local declines and because of widespread decline in the United States.

Great Plains Toad

Bufo cognatus

Lying in suspended animation below ground most of the year is a way of life for these toads, but like the dormant seeds of prairie grasses, Great Plains Toads awaken with the rains and push their way up to the grassland above in search of shallow pools. Abandoning their self-imposed but necessary 10-month period of silence, male Great Plains Toads scream unlike any other North American animal. Peaceful prairies are pierced by pulsing trills that can be heard over two kilometres (1.2 miles) away and that resonate at frequencies so high that listeners often resort to earplugs. • Despite the spirited cacophony of male Great Plains Toads, few ears hear it and fewer still recognize the significance of these

SIMILAR SPECIES

Canadian Toad, p. 168

Plains Spadefoot, p. 158

trills. A short breeding and courtship season for this toad, coupled with a very sparse Canadian population, make encounters with Great Plains Toads rare and truly special events. • The plough has forever altered much of the natural grassland region across Canada's prairie provinces. This dramatic change has resulted in a steep decline in available habitat for animals like the Great Plains Toad, which depend upon native prairie for a home. Thankfully, stewardship of native grasslands by ranchers and federal agencies has maintained significant portions of prairie during the past hundred years, where Great Plains Toads now find their best habitat.

ID: light brown to olive green with dark splotches and more than 5 small "warts" within each splotch; cranial crests form an "L" around inside and back of each eye, almost touching the parotid gland, and form a "V" above nose; small tympanum; white belly.

LENGTH: up to 11 cm ($4^1/_2$").

DISTRIBUTION: from southeastern Alberta to southwestern Saskatchewan. *Selected Sites:* Suffield NWA (AB), Grasslands NP (SK).

HABITAT: native short-grass prairie, particularly sandy areas near water.

ACTIVITY PATTERNS: chiefly nocturnal on rainy nights between May and August.

REPRODUCTION: in response to spring rains, toads emerge from burrows and migrate to shallow breeding ponds; females lay up to 20,000 eggs in long strings; eggs hatch in 2 days and tadpoles transform in 6 weeks, but they may not mature for 3–5 years.

FOOD: *Adult:* primarily invertebrates such as moths, flies and beetles. *Tadpole:* plants and algae.

CALL: a loud, long trill similar to the screech of a slipping fan belt on a car; also likened to a pneumatic drill.

SIMILAR SPECIES: *Canadian Toad* (p. 168): grey spots on belly; cranial crests do not run behind eye; dark splotches usually have 5 or fewer large "warts." *Plains Spadefoot* (p. 158): vertical pupils; single bump between eyes.

FRENCH NAME: Crapaud des steppes

DID YOU KNOW? Great Plains Toads raise themselves up on their legs, lower their heads and inflate their bodies when threatened. Great Plains Toads are listed as Special Concern by COSEWIC.

Canadian Toad

Bufo hemiophrys

This toad's name is close enough to the truth, with its distribution stretching across the prairie provinces and only slightly dipping into just a few areas of the Great Plains states. The Canadian Toad is a fine representative of Canada, living within dry prairie, aspen parkland or boreal forest. • Until recently, the Canadian Toad was quite easy to find or hear over much of its range during the peak of spring breeding. Unfortunately, a yet-to-be-explained decline has occurred in places where this animal was once found. Drought and habitat loss are suspected to be the culprits. These declines could be considered even more alarming to Canadians because this toad has the great majority of its numbers in Canada. • The Canadian Toad has a narrow timeframe in which to be active because for much of the year within its distribution the weather is either too hot or too cold. In order to cope, it remains in burrows or hidden beneath debris for long periods, awaiting appropriate temperatures to prompt it into action.

SIMILAR SPECIES

Great Plains Toad, p. 166

Plains Spadefoot, p. 158

American Toad, p. 162

ID: varies from brown to grey-green or reddish with dark, raised bumps surrounded by black spots (less than 5 bumps per spot); often a light line down middle of back; belly is light with dark spots; cranial crest joins and runs parallel between eyes; large, kidney-shaped parotid glands; tympanum smaller than eye.

LENGTH: 5–7 cm (2–3").

DISTRIBUTION: from the eastern half of Alberta to the western half of Manitoba and north to Fort Smith, Northwest Territories; less commonly found in the North. *Selected Sites:* Riding Mountain NP (MB), Grasslands NP (SK).

HABITAT: borders of ponds and lakes from the grasslands to the boreal forest; breeding and larval development occur in shallow areas of lakes or sloughs or in temporary puddles.

ACTIVITY PATTERNS: depending on the time of year and temperature, active during day or night; seeks shade underground if too hot; hibernates underground.

REPRODUCTION: breeding and courtship occur from May to June; female can lay up to 7000 eggs in single strands; eggs hatch in 3–12 days and tadpoles transform 6–7 weeks later.

FOOD: *Adult:* worms, beetles and ants. *Tadpole:* algae and plants.

CALL: a brief, harsh trill that lasts 2–5 seconds; repeated every 15–20 seconds.

SIMILAR SPECIES: *Great Plains Toad* (p. 166): V-shaped cranial crest runs between eye and parotid gland; more than 5 "warts" in large, dark blotches on back and sides. *Plains Spadefoot* (p. 158): vertical pupils; single bump between eyes. *American Toad* (p. 162): central Manitoba; converging cranial crests; elongated parotid gland. *Western Toad* (p. 164): northern Alberta; no cranial crests.

FRENCH NAME: Crapaud du Canada

SIMILAR SPECIES

Western Toad, p. 164

DID YOU KNOW? *Hemiophrys* is derived from Greek words meaning "half" and "eyebrow." This name aptly describes the Canadian Toad's cranial crest, which is shorter than that of the American Toad and Great Plains Toad.

Fowler's Toad
Bufo fowleri

To most Canadians, the north shore of Lake Erie is known as the tomato hotbed of Canada. Birders, in contrast, know well the virtues of the spring songbird migration at places such as Point Pelee, Rondeau and Long Point. But for the herpetologist, this location is Canada's mecca for the seldom seen Fowler's Toad. • Superficially, there is little to differentiate Fowler's Toad from its much more common and abundant counterpart, the American Toad. There are enough physical characteristics if you see one close-up, but even these are a little tedious to go by. Habitat is the best identifier of Fowler's Toad. Throughout its range, this toad is an animal with an eye for beach and sand. The north shore of Lake Erie is regularly ravaged by storms. Waves and wind scour the shoreline clean, and ice often shaves the lake's perimeter clean of trees. These violent

SIMILAR SPECIES

American Toad, p. 162

changes set the vegetation's ecological clock back a generation or two, which pleases Fowler's Toad. Early successional stages free of older vegetation are preferred by this inconspicuous amphibian.

ID: brown, green or yellowish body; large, dark blotches that usually contain at least 3 "warts"; light stripe down centre of back; long parotid glands run along and meet prominent cranial crests; crests are L-shaped and do not meet above nose; belly is usually unmarked or with a single spot. *Male:* dark throat. *Female:* light throat.

LENGTH: 4–6 cm (1½–2½").

DISTRIBUTION: in Canada, limited to a few spots along the north shore of Lake Erie and south shore of Lake Ontario. *Selected Sites:* Long Point PP, Rondeau PP (both in ON).

HABITAT: sandy, sparsely vegetated shoreline and lakeside marshes.

ACTIVITY PATTERNS: generally active from April to September with a peak in activity in May and June; most active on rainy nights, but can be encountered during the day as well.

REPRODUCTION: breeds in April; may breed in communal ponds; up to 6000 eggs are laid in strings; tadpoles transform in late June or early July; sexually mature in 2 years.

FOOD: *Adult:* aquatic and terrestrial invertebrates. *Tadpole:* plants and algae.

CALL: has been compared to the sound of a baby's cry or sheep baying; a short nasal *waaa* sound lasting 1–3 seconds.

SIMILAR SPECIES: *American Toad* (p. 162): parotid glands do not touch cranial crests; mottled belly; may hybridize with Fowler's Toad.

FRENCH NAME: Crapaud de Fowler

DID YOU KNOW? Fowler's Toad was named after S.P. Fowler, a naturalist who lived and worked among these toads along the Massachusetts shore. Fowler's Toad is listed as Threatened by COSEWIC.

Red-legged Frog

Rana aurora

A red flag has been raised concerning the state of the Red-legged Frog. Over most of its limited range in Canada, this sharp-looking frog has disappeared or declined in many of its historic breeding sites. The causes are all the usual ones, with ponds being filled in and paved over, polluted with agricultural run-off or overrun by non-native intruders. This is a tired old story in biological conservation, but one that continues to plague populations of threatened animals. • The introduction of the American Bullfrog to Vancouver Island and the Lower Mainland has been particularly tough on the Red-legged Frog. Historically, the Red-legged Frog alone occupied the large frog niche, living comfortably alongside the occasional salamander, toad and newt. With the coming of the Bullfrog, the balance has been tipped, with the upper hand going to the Bullfrog. Ponds where the two are found play out the "eat or be eaten"

SIMILAR SPECIES

Columbia Spotted Frog, p. 184

Oregon Spotted Frog, p. 182

Wood Frog, p. 188

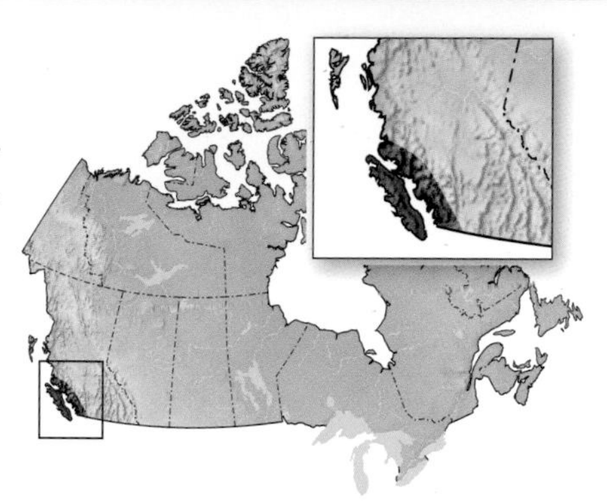

rule of nature, with the Bullfrog usually winning out. As a consequence, the Red-legged Frog is generally now found only in areas that have not yet been colonized by the Bullfrog. That said, a journey off the beaten path is all that's required to find one of these animals. It tends to be fairly terrestrial, so it's not unusual to discover one as it jumps away from under your foot as you stroll along a shoreline trail.

ID: rust-coloured to grey body with small, dark markings; distinct dorsolateral ridges; light stripe on jaw bordering a dark "mask;" chest is white to light yellow; belly and inner thighs are red, with red becoming more intense with age; green mottling in groin area; small tympanum; toes are not fully webbed.

LENGTH: 5–12 cm (2–5").

DISTRIBUTION: the Lower Mainland, Fraser Valley, Vancouver Island and other Gulf Islands; has been introduced to Haida Gwai. *Selected Sites:* Haida Gwai especially around Port Clements, Gulf Islands, southwestern Vancouver Island (all in BC).

HABITAT: forests that have permanent but small ponds; habitually travels overland in the shade of cedars and hemlocks during summer.

ACTIVITY PATTERNS: comparatively long period of activity due to the mild West Coast climate; most active at night and sometimes on overcast days.

REPRODUCTION: courting male begins vocalizing in late February or March; female lays a mass of 500–1000 large eggs on the stems of aquatic vegetation; tadpoles transform during summer; mature in 2–3 years.

FOOD: *Adult:* invertebrates such as worms, beetles, spiders and flies. *Tadpole:* plants and algae.

CALL: weak call; paraphrased as *uh uh uH UH*; lasts about 2 seconds and does not carry far, as it is often delivered under water.

SIMILAR SPECIES: *Columbia Spotted Frog* (p. 184) and *Oregon Spotted Frog* (p. 182): upward-directed eyes; when viewed from above more eye than eyelid; lack dark "mask;" toes are fully webbed; spots tend to be larger with light centres. *Wood Frog* (p. 188): rarely has red on belly or legs.

FRENCH NAME: Grenouille à pattes rouges

DID YOU KNOW? *Aurora* means "dawn" or "light," referring most likely to the bright red legs of this species. The Red-legged Frog is listed as Special Concern by COSEWIC.

American Bullfrog

Rana catesbeiana

American Bullfrogs are highly aquatic eating machines that are doing well in most parts of their natural and introduced range in Canada. Introduced to parts of British Columbia by frog-leg farming entrepreneurs, these amphibians have since escaped the confines of that failed business and have set up in the native lakes and ponds of the area. • Their aggressiveness has been a major setback in the population of native frogs not only along Canada's West Coast, but also wherever in the world American Bullfrogs have been placed. They continue to be a fairly common and pleasing sight in their natural habitat from Ontario to the Maritimes. In some places, American Bullfrogs are finding the going a little rough as a result of acidification, habitat loss and diseases, but for the most part the Canadian status of the American Bullfrog remains stable.

SIMILAR SPECIES

Green Frog, p. 176

ID: colour ranges from light green to dark brown on back and light underneath with small, dark markings; dorsolateral ridges wrap around large tympanum and do not extend along back; tadpoles can be very large in comparison to other species. *Male:* tympanum larger than eye. *Female:* tympanum same diameter as eye. *Breeding male:* yellow chin.

LENGTH: up to 20 cm (8").

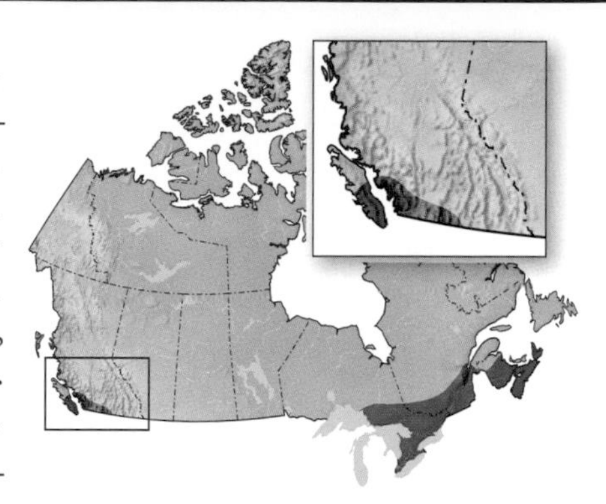

DISTRIBUTION: naturally occur from southern Ontario to Nova Scotia; introduced to the Lower Mainland and Vancouver Island in British Columbia; widely introduced all over the world. *Selected Sites:* Algonquin PP (ON), La Mauricie NP (QC), Fundy NP (NB), Kejimkujik NP (NS).

HABITAT: need permanent water bodies for egg-laying and larval development; lakes, large ponds and reservoirs with well-vegetated shorelines are preferred.

ACTIVITY PATTERNS: active from late spring to mid-fall; courtship often occurs during warm nights; can be seen sunning themselves and feeding during the day.

REPRODUCTION: initiate courtship and breeding activities in mid-June to late July; males determine territories and call to females, who choose a mate based on the territory and on the size and age of the male, which they determine by the pitch of his call; females lay large egg masses of up to 8000 eggs that spread out over the water's surface; tadpoles become large and do not transform until their second or third year; males mature 1 year after transformation; females mature 2 years after.

FOOD: *Adult:* insects, fish, other amphibians, small birds and mammals. *Tadpole:* algae and vegetation.

CALL: deep, monotone *jug-o-rum more-rum*. In Quebec, the onomatopoeic *ouaouaron* is preferred.

SIMILAR SPECIES: *Green Frog* (p. 176): smaller; 2 dorsolateral ridges run along each side of back.

FRENCH NAME: Ouaouaron

Green Frog

Rana clamitans

Although most frogs are green, this is the only one to wear the description in its name. The Green Frog is a widespread and common amphibian over much of the eastern half of Canada, and it is one of the best candidates to be the first frog someone catches. Unfortunately, it suffers from a bit of an identity crisis owing to its similarity to the much better known American Bullfrog, with whom it is often found. • The Green Frog's song definitely doesn't live up to the stereotypical vocalization of frogs. It doesn't croak or grunt, and it certainly doesn't "ribbit." Its banjo-like twang is about as off beat as you'd expect, but though it may be the best-known sound of the Green Frog, it is not the only one. Frogs, like birds and humans, are vocal animals that have different sounds for different situations. Leaping away from danger, the Green Frog makes a loud *reep*. And

SIMILAR SPECIES

American Bullfrog, p. 174

Mink Frog, p. 186

Wood Frog, p. 188

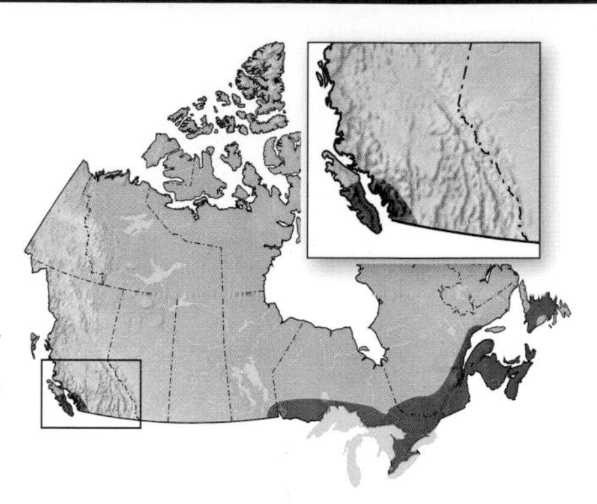

if a male suffering from gender confusion grasps other courting males, yet another call is produced!

ID: green, bronze, blue or brown body; prominent dorsolateral ridges along entire length of torso; upper lip is usually green; belly is white with darker markings; hind legs have dark bars perpendicular to axis of leg. *Male:* yellow throat; tympanum twice the size of eye. *Female:* creamy throat; tympanum same diameter as eye.

LENGTH: 7–10 cm (3–4").

DISTRIBUTION: from Manitoba to all three Maritime provinces; introduced to Newfoundland and British Columbia. *Selected Sites:* Fundy NP (NB), Kejimkujik NP (NS), Algonquin PP (ON).

HABITAT: in and along edges of permanent water such as lakes, swamps and streams.

ACTIVITY PATTERNS: becomes active later in the year than other frogs; tends to be most active from mid-afternoon through the night, but can be found moving about at other times as well.

REPRODUCTION: males have breeding territories; breed from early summer through to August; females lay eggs in a mass on the water surface; eggs hatch within 3 weeks; tadpoles tend to be in deep water around aquatic vegetation; tadpoles overwinter in water and transform the next spring.

FOOD: *Adult:* invertebrates; larger individuals might also eat small fishes or amphibians. *Tadpole:* plants and algae.

CALL: banjo-like *twang*, usually only once but sometimes repeatedly; usually calls from under bushes at the edge of its pond or lake.

SIMILAR SPECIES: *American Bullfrog* (p. 174): ridge runs around tympanum; no dorsolateral ridges. *Mink Frog* (p. 186): yellowish belly; upturned eyes; musky odour when rubbed; dark, irregular blotches, not transverse bars, on hind legs. *Wood Frog* (p. 188): dark "mask."

FRENCH NAME: Grenouille verte

DID YOU KNOW? Recent studies on the Green Frog in Quebec and Ontario have found that severe deformities such as extra limbs are linked to the use of pesticides in orchards and other agricultural industries.

Pickerel Frog

Rana palustris

If the pattern on the Northern Leopard Frog looks to be designed by an artist, then the square, symmetrical spots of the Pickerel Frog look to be designed by an engineer. The Pickerel Frog has a more regulated appearance to it, but can still be mistaken for its much more widespread counterpart. The Pickerel Frog is known to interbreed with the Northern Leopard Frog, resulting in some confusing patterns. However, the Pickerel Frog is yellow or orange under the hind legs and on the groin. • All frogs in Canada are closely tied to water, but only this one has the Latin word for marshes, *palustris,* tied to it. Even the English name does not refer specifically to the frog, but to its use as bait when fishing for pickerel, otherwise known as Walleye. Unfortunately for the fish, the Pickerel Frog secretes an irritating toxin from its skin. Natural predators of the Pickerel Frog include American Bullfrogs and Green Frogs, both of which have developed a resistance to the toxin.

SIMILAR SPECIES

Northern Leopard Frog, p. 180

DID YOU KNOW? Although the Pickerel Frog is reputed to have strong toxins in its skin and is avoided by many predators, it is also eaten by those same predators. Perhaps the toxic reputation of this attractive frog needs to be reassessed.

ID: light brown, smooth skin; large, yellow or white dorsolateral ridges; yellow belly; yellow undersides of hind legs; dark, square spots surrounded by light borders run down back.

LENGTH: 4–9 cm (1½–3½").

DISTRIBUTION: southern Ontario, southern Quebec, New Brunswick and Nova Scotia; more common in the Maritimes and less common as the range extends west. *Selected Sites:* Kejimkujik NP (NS), McCabe Lake in Halifax County (NS), Mono Cliffs PP (ON), Gould Lake Conservation Area (ON).

HABITAT: springs, seepages, ponds and streams.

ACTIVITY PATTERNS: after breeding, will spend part of summer foraging in fields and ditches near water.

REPRODUCTION: breeding begins in mid- to late spring; 1000–3000 eggs are laid in one or more globular clumps attached to vegetation in still water; eggs hatch in 11–21 days; tadpoles transform after about 80 days; take 2–3 years to reach maturity; typically live 4 years.

FOOD: aquatic invertebrates such as snails; terrestrial invertebrates.

CALL: a quiet snore that does not carry very far; similar to the mooing of a cow; may call from under water; aggressive male will give a raspy chuckle.

SIMILAR SPECIES: *Northern Leopard Frog* (p. 180): can be green or brown; spots are more round or oval; belly and underside of hind legs are creamy.

FRENCH NAME: Grenouille des marais

Northern Leopard Frog

Rana pipiens

Ask someone to illustrate a frog, and they'll almost certainly draw a Northern Leopard Frog. The combination of green body and black spots seems to be ingrained in most people's minds when they visualize a frog, which seems pretty fair because the Leopard Frog, next to the Wood Frog, is the most widespread, easily encountered species in Canada, not to mention that it is the frog most dissected in biology classes throughout North America. • The Northern Leopard Frog is particularly common in central and eastern Canada, where it can be expected in most wetlands that meet its habitat requirements. Surviving winter is one of the most challenging life obstacles for most Canadian amphibians, and the Northern Leopard Frog is no exception. It tends to remain submerged in wetlands in winter. The cold, but not freezing, water allows the frog to slow its metabolism to the cusp of death, neither acquiring nor expending much-valued energy. • During the mid- to late 1970s, this large, vocal and prominent Canadian frog experienced localized extinctions that were more extreme and widespread as one went west from Ontario. The reasons for the decline remain simply guesses. In some areas populations have recovered, but elsewhere, especially in British Columbia, the Northern Leopard Frog has not recovered. Although some amphibians are well

SIMILAR SPECIES

Pickerel Frog, p. 178

Columbia Spotted Frog, p. 184

known to be in decline, it is alarming that one of the most conspicuous frogs in Canada should also be in decline, particularly because there is no explanation for it.

ID: green or light brown body; dark, irregular spots edged with light borders; obvious dorsolateral ridges; whitish belly.

LENGTH: 4–10 cm (1½–4").

DISTRIBUTION: southeastern British Columbia and south-central Northwest Territories; across the prairies and eastern Canada including Labrador; introduced in Newfoundland; not found in the Yukon *Selected Sites:* La Mauricie NP (QC), farm ponds throughout southern Ontario and Quebec.

HABITAT: along streams, lakes, beaver ponds, oxbows, sloughs and permanent wetlands; during summer is encountered in fields or grassy meadows.

ACTIVITY PATTERNS: active during the day, from spring through fall.

REPRODUCTION: breeding occurs in mid-spring; female attaches her egg masses to submerged vegetation; softball-sized egg masses usually contain up to 3500 eggs; as with all aquatic amphibians, timing of hatching and transformation is very dependent on water temperature; tadpoles transform in approximately 6 weeks; mature in approximately 2 years.

FOOD: *Adult:* small invertebrates. *Tadpole:* plants and algae.

CALL: courtship call sounds like one snore followed by soft grunting; gives the impression of a slow, sleepy animal.

SIMILAR SPECIES: Eastern Canada: *Pickerel Frog* (p. 178): regular, squared spots; yellow belly; brown body. Western Canada: *Columbia Spotted Frog* (p. 184): smaller black spots with no halos; brownish body.

FRENCH NAME: Grenouille léopard

DID YOU KNOW? Extirpated populations in western Canada are now being experimentally re-populated by relocating frogs from wild populations. The Southern Mountains population is listed as Endangered, and the Western Boreal/Prairies population as Special Concern, by COSEWIC. More eastern populations have declined but are still deemed Not at Risk.

Oregon Spotted Frog

Rana pretiosa

Oregon Spotted Frogs have an interesting breeding process. As a female works her way toward her targeted mate (the biggest and loudest), males weaker in voice or stature strive to intercept her and piggyback atop her (known as amplexus), and are occasionally successful in fertilizing her eggs. In most cases, though, the prime males dislodge the smaller interlopers. • Canada's first discovered population of Oregon Spotted Frogs is now extinct, succumbing to a variety of pressures including French cuisine and bass fishermen in search of bait. Fortunately, a second population of these handsome frogs has recently been discovered in the extreme southwestern corner of British Columbia. In Washington and Oregon, similar

SIMILAR SPECIES

Red-legged Frog, p. 172

Columbia Spotted Frog, p. 184

threats to these frogs have had equally disturbing effects on the population.

• Oregon Spotted Frogs are much more aquatic than other Canadian frogs and are therefore difficult to find. When disturbed, they will dive down to the bottom of their pond and stay there for quite some time. They only leave the water for short periods when foraging, and they never travel between ponds except by connecting waterways. As such, they are especially vulnerable to fragmentation of their habitat.

ID: dark green to light brown base colour; irregular spots, often with light centres; light stripe on upper jaw; slightly upturned eyes; obvious tympanum; belly may be yellow, orange or red, and these colours are often more vibrant on inner thighs.

LENGTH: 7–8 cm (3").

DISTRIBUTION: the lower Fraser Valley in southwestern British Columbia. *Selected Sites:* anywhere within their range and habitat.

HABITAT: permanent ponds, sluggish streams or lakes that are bordered by non-woody vegetation.

ACTIVITY PATTERNS: have a long period of activity because of the mild West Coast climate; generally most active at night, but can also be active by day.

REPRODUCTION: breed in early spring with female laying a single egg mass containing 700–1500 individual eggs; eggs hatch in approximately 4 days; tadpoles normally transform by the end of summer; take 2–3 years to reach sexual maturity.

FOOD: *Adult:* primarily aquatic invertebrates, but occasionally terrestrial invertebrates. *Tadpole:* aquatic plants and algae.

CALL: rapid series of short clucking notes building in volume to the end; up to 10 seconds long; calls are low-pitched and often made under water, so can only be heard close by.

SIMILAR SPECIES: *Red-legged Frog* (p. 172): incomplete webbing on toes; eyes are not upturned; spotting on back is usually smaller; inner thighs are yellowish red. *Columbia Spotted Frog* (p. 184): larger head; no mottling at throat.

FRENCH NAME: Grenouille maculée de l'Orégon

DID YOU KNOW? Oregon Spotted Frog eggs are very sensitive to heat; thus breeding ponds require very stable water temperatures. Oregon Spotted Frogs are listed as Endangered by COSEWIC.

Columbia Spotted Frog

Rana luteiventris

The Columbia Spotted Frog never seems to venture more than a leap or two from water. Its highly adapted body reflects that of a water-lover—its feet are fully webbed for efficient paddling, and its eyes are placed high on the head. The reason for this placement is obvious when you see one staring at you with only its globular eyes popping above the water's surface and its body safely under water.

The Columbia Spotted Frog doesn't jump all that well in comparison to similar-sized frogs, but it swims fluidly through water with grace normally reserved for fish. • The Columbia Spotted Frog also has an intriguing behaviour that is more noted in ostriches than in frogs. A frog spooked from the shoreline is very often observed swimming panic-stricken to the bottom. Once it reaches

SIMILAR SPECIES

Red-legged Frog, p. 172

Oregon Spotted Frog, p. 182

Northern Leopard Frog, p. 180

the bottom, the frog gives a few kicks and buries itself in mud, but often only neck deep. With the body and hind legs exposed for all to see, the effectiveness of the frog's "see no evil" defensive strategy is questionable, but humourous nonetheless.

ID: dark green to light brown base colour; irregular spots often with light centres; light stripe on upper jaw; upturned eyes; belly may be yellow, orange or red and darker on inner thighs; toes are completely webbed.

LENGTH: 5–9 cm (2–3½").

DISTRIBUTION: mountains of Alberta and eastern British Columbia and barely into the southern Yukon Territory; declining in Alberta owing to development in the mountain areas. *Selected Sites:* Kananaskis Country (AB), Rocky Mountain parks (BC and AB).

HABITAT: permanent lakes and ponds in montane and subalpine ecosystems within mixed-wood forests.

ACTIVITY PATTERNS: active at just about any time day or night; intensity of courting activity increases during the evening and nighttime hours; becomes active in early spring until mid-fall.

REPRODUCTION: breeds in early spring; females lay egg masses containing 700–1500 eggs each in communal bunches; eggs hatch in about 4 days; tadpoles either transform by the end of summer or they overwinter; reach reproductive age at 4–5 years.

FOOD: *Adult:* many types of aquatic invertebrates; occasionally terrestrial invertebrates. *Tadpole:* plants and algae.

CALL: a series of quick, whispery grunts that grow louder throughout the call but are still weak in volume; sounds do not carry a great distance.

SIMILAR SPECIES: *Red-legged Frog* (p. 172): toes not fully webbed; eyes lower on head; spotting on back is usually smaller; groin region has yellowish green patches. *Oregon Spotted Frog* (p. 182): smaller head; mottling at throat. *Northern Leopard Frog* (p. 180): very distinct dark spots with light halos.

FRENCH NAME: Grenouille maculée de Columbia

DID YOU KNOW? Until recently, the Columbia Spotted Frog was simply known as the "Spotted Frog" (*R. pretiosa*), the scientific name that now refers only to the Oregon Spotted Frog.

Mink Frog

Rana septentrionalis

The Mink Frog is your typical frog on a lilypad, using the flat green surfaces as an escape route to deeper water, for camouflage and cover or for a hunting platform. Shy and reclusive, the Mink Frog will reveal its presence in the summer months with its loud, knocking mating call, reminiscent of two stones being banged together. The Mink Frog usually remains close to a water body, but rainy days may bring it inland, especially if the nearby bogs or ponds are joined together by a cool, moist forested area. • The Mink Frog's eyes are placed higher up on its head compared to other frogs, perhaps so it can scout out potential insect meals hovering above or watch for avian predators that would like to lunch on a frog. If a predator does get its mouth around a Mink Frog, however, it is in for an unsavory surprise. The frog releases a mildly stinky skin secretion when it is handled. It supposedly resembles the smell of a mink, but as most people are unfamiliar with mink musk, think of a faint smell of rotting onions.

SIMILAR SPECIES

Green Frog, p. 176

ID: olive green or brown with dark blotches over entire body including legs; may become more green toward head and especially mouth; underside is cream coloured; eyes are placed high on head; webbing extends to last joint on longest toe; may or may not have dorsolateral ridges. *Male:* tympanum is larger than eye; 2 vocal sacs. *Female:* tympanum is same size as eye or smaller.

LENGTH: 5–8 cm (2–3").

DISTRIBUTION: from southeastern Manitoba through southern and central Ontario and Quebec and into the Maritimes and Labrador. *Selected Sites:* Algonquin PP (ON), La Mauricie NP (QC), Kejimkujik NP (NS).

HABITAT: very slow-moving or still water bodies; pond, marsh or lake edges with heavy aquatic vegetation, especially lily pads and water shield.

ACTIVITY PATTERNS: active during the day; burrows into mud and lies dormant during winter.

REPRODUCTION: male begins to sing in late June through July; female lays up to 4000 eggs per season; a loose mass of eggs about the size of a tennis ball is released and attached to aquatic vegetation; eggs hatch after 1 week; the young remain as tadpoles until the next summer; mature 1–2 years after transformation.

FOOD: both terrestrial and aquatic invertebrates; beetles, 2-winged flies, ants and aquatic insect larvae.

CALL: male's resonant mating call is a raspy, staccato *tuk tuk tuk;* if many males are calling simultaneously, the noise can sound like horse hooves on pavement.

SIMILAR SPECIES: *Green Frog* (p. 176): less black mottling on body; dark markings occur as bands on legs; white belly; webbing extends to second last joint on longest toe.

FRENCH NAME: Grenouille du nord

DID YOU KNOW? The Mink Frog is the most Canadian of frogs, as most of its range is in Canada. It is the only native frog that smells distinctively to people.

Wood Frog
Rana sylvatica

Wood Frogs couldn't be more Canadian if they wore RCMP stetsons and ate poutine. They are the most northerly occurring amphibians in the world, occupying boreal forest in every province and territory of Canada. Like all amphibians and reptiles, Wood Frogs have very limited abilities to produce body heat, and their activity is generally determined by ambient temperature. However, these frogs, like most Canadians, are tolerant of the bitterly cold winters that define this country. • Canadian researchers were the first to explain Wood Frogs' resistance to the cold. When temperatures decline to near freezing, this species can produce complex sugars and proteins that draw the water from inside cells to the area surrounding the cells. Therefore, when the cells freeze they will not rupture from expanding

SIMILAR SPECIES

Columbia Spotted Frog, p. 184

Boreal Chorus Frog, p. 200

Green Frog, p. 176

water. Simply put, these frogs can survive being frozen. That frogs can freeze is not such a big deal, but frogs that can freeze solid and then live to jump again, that's a great trick. But even Wood Frogs have limits, and the specific area in which they hibernate must never drop below -3° C (25° F). Protection from intolerably cold temperatures is found under logs or leaf litter on the forest floor, where the frogs can bide their time waiting for warmer weather. • Spring is a time of celebration, and hardy Wood Frogs are among the first frogs to take to thawing ponds in search of love. Their call is not the most persistent, but the hoarse cackling sounds and stirring of pond sediments is an undeniable sign of spring, even if the ground remains dusted with winter white.

ID: tan to dark body; "mask" covering eye and ear area; some individuals have a white stripe down back; underside is lighter than back colour; conspicuous ridges or folds along back.

LENGTH: 5–8 cm (2–3").

DISTRIBUTION: every province and territory; have been successfully introduced to Newfoundland. *Selected Sites:* Cape Breton Highlands NP (NS), La Mauricie NP (QC), Algonquin PP (ON), Pukaskwa NP (ON), Riding Mountain NP (MB), Prince Albert NP (SK), Elk Island NP (AB), Wood Buffalo NP (AB and NT).

HABITAT: woodlands, beaver ponds, semi-permanent wetlands and small lakes; may occur in tundra or grasslands.

ACTIVITY PATTERNS: active from early spring through fall, primarily during the day but also during warm nights; hibernate in organic litter during winter.

REPRODUCTION: early breeders; egg mass the size of a tennis ball rests just below the water surface, attached to submerged vegetation; populations may lay clusters together, resulting in large egg masses; tadpoles transform after 1–3 months.

FOOD: *Adult:* insects, spiders, worms and other invertebrates; lunge with open mouths to catch prey. *Tadpole:* plants and algae.

CALL: courtship call consists of short, grunty cackling; duck-like.

SIMILAR SPECIES: *Columbia Spotted Frog* (p. 184): colourful underside. *Boreal Chorus Frog* (p. 200): very small; 3 stripes down back; sticky toe pads. *Green Frog* (p. 176): no dark "mask."

FRENCH NAME: Grenouille des bois

DIDYOU KNOW? Wood Frogs' adaptation to northern climates interests researchers working on developing more effective techniques for storing organs intended for human transplant and on diabetes research.

Northern Cricket Frog

Acris crepitans

The Northern Cricket Frog's only toehold in Canada is Pelee Island, and it is almost certainly extirpated there, too. It's a shame that this tiny frog is not more widely distributed in Canada, as its chirps delightfully complement the summer symphonies of crickets during the evenings. It is this insect impersonation that inspired this frog's name (*Acris* also refers to crickets and grasshoppers), but its similarities with invertebrates run deeper than a mere title. Like insects, this pint-sized amphibian is short lived; even an old-timer rarely celebrates its third birthday. The Northern Cricket Frog is also very small, so tiny that many kinds of invertebrates are much larger than it, and even reverse the links of the food chain by feeding on this vertebrate. • The Northern Cricket Frog has never been widespread in Canada, but recent

SIMILAR SPECIES

Spring Peeper, p. 198

Western Chorus Frog, p. 202

Gray Treefrog, p. 194

declines have earned it a place on our endangered species list. Habitat loss and pesticide use have been implicated in declines elsewhere and have undoubtedly been contributing factors for losses in Canada. This frog has not been reported on the mainland since 1972, and on Pelee Island an increase in the number of American Bullfrogs is thought to have been the cause of further declines in numbers of Northern Cricket Frogs. American Bullfrogs have no respect for family bloodlines and gladly snack on the tiny frog when the opportunity presents itself. It is now likely that there are no longer any Northern Cricket Frogs in Canada. Objectives have been formulated and some action has been taken to return this animal to sites on Pelee Island and Point Pelee National Park.

ID: body colour ranges from yellow to brownish-green to red or black; "warty" skin; dark triangle between eyes; short legs.

LENGTH: up to 4 cm (1½").

DISTRIBUTION: last reported (but not confirmed) in Canada on Pelee Island in 1987; historically found in and around Point Pelee NP (ON).

HABITAT: marshes, permanent wetlands, ditches and along streams; takes shelter in thick emergent vegetation; open areas are used for basking.

ACTIVITY PATTERNS: active from late spring to early fall; most active on warm nights; can be heard on warm, rainy days.

REPRODUCTION: late breeders; court and lay eggs in mid-summer; females lay up to 400 eggs that are placed individually on vegetation or deposited on the bottom of a pond; eggs hatch in 3–4 days; full transformation occurs in 5–10 weeks.

FOOD: *Adult:* small insects and other invertebrates.

CALL: breeding call is a persistent *kick-kick-kick*, like 2 small pebbles being clicked together; similar to a cricket.

SIMILAR SPECIES: *Spring Peeper* (p. 198): dark "X" on back. *Western Chorus Frog* (p. 202): 3 dark lines running down back. *Gray Treefrog* (p. 194): orange-yellow inner thighs; light spot with a dark halo under each eye.

FRENCH NAME: Rainette grillon

DID YOU KNOW? The Northern Cricket Frog is capable of jumping over one metre (40 inches). The subspecies that once called Canada home was known as Blanchard's Cricket Frog. The Northern Cricket Frog is listed as Endangered by COSEWIC, although no reliable record of it exists over the past 20 years.

Cope's Gray Treefrog

Hyla chrysoscelis

Cope's Gray Treefrog looks identical to the Gray Treefrog. Most books assure readers that these twins of the Treefrog world can be separated by voice alone. This is true enough—the trill of the Cope's is faster and more buzzy than musical. Although the speed of the trill varies according to temperature,when the two species call together they can still be differentiated. • Through genetic testing, science has shared with us the mystery of the treefrogs' confusion. Both are genetically similar except in the number of chromosomes. Cope's Gray Treefrog has the usual two sets of chromosomes. The Gray Treefrog is known as a "tetraploid" and has four sets of chromosomes. • All of this detail matters little to people sitting on

SIMILAR SPECIES

Gray Treefrog, p. 194

Spring Peeper, p. 198

Boreal Chorus Frog, p. 200

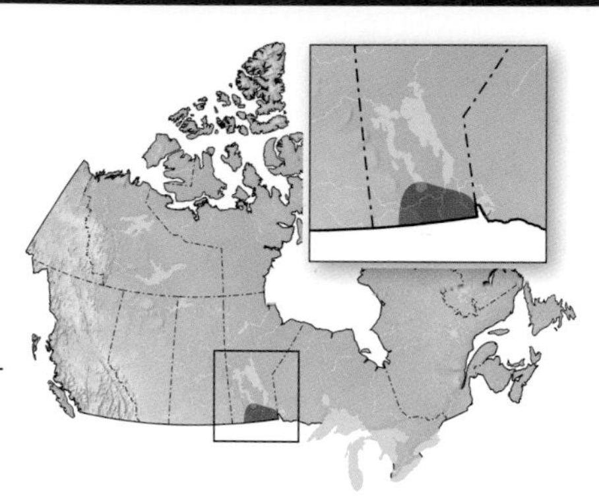

their porches enjoying the sound of the puttering trill, whether it be fast or slow. Both species can set up deafening yet pleasant choruses when they congregate at a pond to breed. They also tend to call outside the breeding season, even into late fall when nights are warm and humid with drizzle.

ID: "warty" with green, brown or grey skin; large, darker marks on back; enlarged toe pads; dark-rimmed light spot below each eye; underside of thighs is yellow-orange.

LENGTH: up to 6 cm ($2^1/_2$").

DISTRIBUTION: southeastern Manitoba. *Selected Sites:* Whiteshell PP (MB).

HABITAT: trees and shrubs standing near permanent water; older trees are preferred, but these frogs can also be found on buildings, roads, fences and other areas when it is rainy or humid.

ACTIVITY PATTERNS: active from late spring to early fall; tend to be most active on warm nights but can be found regularly on warm rainy days; will hibernate under snow cover and leaf litter; can freeze solid.

REPRODUCTION: in Canada, courtship and breeding occur from late spring to early summer; on warm evenings, males call from trees; female approaches a male based on his call, then enters the water with her mate; female lays up to 2000 eggs in clusters of 10–40 eggs; egg masses are attached to underwater vegetation; eggs hatch within 5 days; tadpoles transform in 6–8 weeks.

FOOD: invertebrates foraged in trees or shrubs.

CALL: loud, fast and high trill delivered in bursts of 1–3 seconds.

SIMILAR SPECIES: *Gray Treefrog* (p. 194): lower and slower call. *Spring Peeper* (p. 198): dark "X" on its back; much smaller. *Boreal Chorus Frog* (p. 200): smoother skinned; much reduced toe pads; 3 stripes on back.

FRENCH NAME: Rainette criarde

DID YOU KNOW? While camouflaged skin makes Cope's Gray Treefrog virtually invisible against the bark of trees, it also presses itself tightly against the tree for added suction and in doing so ensures that it does not cast a shadow.

Gray Treefrog
Hyla versicolor

This sticky-toed counterpart to the famed tropical treefrogs climbs and proudly sings from tree branches in spring. With its habit of clinging to tree trunks with its sticky toe pads, the acrobatic tendencies of the Treefrog family are represented in the Gray Treefrog. Unlike the vivid green of many store-bought varieties, the Gray Treefrog is an illusionist. Clinging to vertical surfaces, this frog is most often overlooked as a result of its mottled and muted appearance. To add to its camouflage, the Gray Treefrog has a limited ability to change the colour of its skin to match the background upon which it rests. For this reason, this treefrog is most commonly encountered through ears and not eyes. • Most visual encounters, however, occur on the frog's terms, as individuals crawl up fences, houses and windowpanes. In these instances, its muddy colours offer little concealment against the bright tones of human décor, and it often rests within easy reach. Gently handling a

SIMILAR SPECIES

Cope's Gray Treefrog p. 192

Spring Peeper, p. 198

Boreal Chorus Frog , p. 200

Gray Treefrog is an experience unlike any other, as the sticky toes designed to cling to vegetation are just as effective in keeping the frog in an upright and safe position on your fingers.

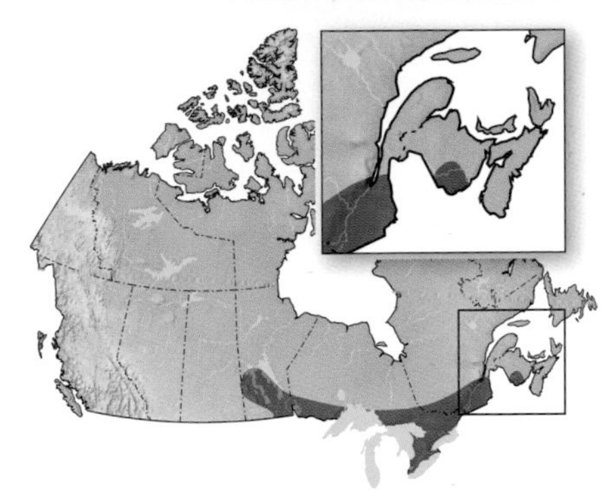

ID: "warty" green, brown or grey skin; large, dark marks on back; enlarged toe pads; dark-edged light spot below each eye; underside of thighs is yellow-orange.

LENGTH: up to 6 cm (2½").

DISTRIBUTION: from eastern Manitoba through to New Brunswick. *Selected Sites:* Algonquin PP, Rondeau PP, Point Pelee NP (all in ON).

HABITAT: trees and shrubs around permanent water; older trees are preferred; can also be found on buildings, roads, fences and other areas when the weather is rainy or humid.

ACTIVITY PATTERNS: active from late spring to early fall; tends to be most active on warm nights, but it can be found regularly on warm, rainy days.

REPRODUCTION: courtship and breeding occur from late spring to early summer; on warm evenings, the males call from trees; female selects a male based on call and territory, then both enter nearby waters to mate; female lays up to 2000 eggs in clusters of 10–40 eggs; masses are attached to underwater vegetation; eggs hatch within 5 days; tadpoles transform 6–8 weeks later.

FOOD: insects including beetles, flies and other invertebrates are foraged in trees or shrubs.

CALL: slow, regular, musical trill easily imitated by vibrating one's tongue on the roof of the mouth.

SIMILAR SPECIES: *Cope's Gray Treefrog* (p. 192): higher and faster call; overlaps only in Manitoba. *Spring Peeper* (p. 198): "X" on back; much smaller. *Boreal Chorus Frog* (p. 200) and *Western Chorus Frog* (p. 202): smoother skin; reduced toe pads; lines down back.

FRENCH NAME: Rainette versicolore

SIMILAR SPECIES

Western Chorus Frog, p. 202

Pacific Treefrog

Hyla regilla

The Pacific Treefrog occurs only in British Columbia, but just about all Canadians are familiar with its voice. Ask any child about the voice of a frog and you will hear the deeply ingrained "ribbit-ribbit." The famed voice of the Pacific Treefrog has sounded in the background of so many TV and film productions that the association between "ribbit" and frog is likely never to be broken. On the West Coast, the Pacific Treefrog tends to stay primarily in the cool and misty forests under the ever-present threat of rain, where it vocalizes on occasion. •Unfortunately for the eager seeker, the Pacific Treefrog is very skilled in the art of concealment. Its sticky toe-pads enable it to climb high in vegetation or nimbly through dense weaves of vines and shrubs. •A small number of tadpoles of the Pacific Treefrog were introduced to Haida Gwai (Queen Charlotte Islands) in the 1960s, and within a few years the moist habitat and lack of predators led to a population explosion. The habitat there is excellent for frogs, but they could not get there without the help of people.

SIMILAR SPECIES

Rocky Mountain Tailed Frog, p. 154

Coastal Tailed Frog, p. 156

Wood Frog, p. 188

ID: small with long toes, each bearing a prominent pad; dark brown or black line running from nose through eye and face to shoulder; often a dark triangle between eyes; skin colour and patterning is variable and unreliable as a field mark.

LENGTH: 3–5 cm (1–2").

DISTRIBUTION: southern British Columbia from Vancouver Island east to the Alberta border; has been introduced to Graham Island and the Queen Charlotte Islands. *Selected Sites:* Vancouver Island, the Lower Mainland, Okanagan Valley (all in BC).

HABITAT: lives on or close to the ground among dense vegetation; mostly vines and shrubs close to water; on occasion it can be found climbing up the sides of homes.

ACTIVITY PATTERNS: can be heard calling at just about any time of day, though it tends to be most active from mid-afternoon through the night; not active during the coldest months of the year.

REPRODUCTION: begins courtship vocalizations in March or April; fertilized female lays small clusters of 20–50 eggs in April or May; tadpoles will transform within 8–10 weeks.

FOOD: *Adult:* small invertebrates. *Tadpole:* plants and algae.

CALL: diagnostic series of 2 short, high-pitched notes; can be heard as *kreck-ek* or *wreck-it*; even several kilometres away.

SIMILAR SPECIES: *Rocky Mountain Tailed Frog* (p. 154) and *Coastal Tailed Frog* (p. 156): vertical pupils; less-defined line through eyes. *Wood Frog* (p. 188): dorsolateral ridges on back; lacks toe pads.

FRENCH NAME: Rainette du Pacifique

Spring Peeper

Pseudacris crucifer

Forget robins, cherry blossoms and prognosticating groundhogs—if you really want to know when spring starts, just listen for Spring Peepers. Their peeping chorus is grand and their simple message clear: spring is here to stay. The gentle and persistent *peeps* are repeated so continuously that the sound fades into the background of wetlands, parks and even backyards for much of spring. Despite weighing little more than a stick of gum, they have a superb gift of producing incredible volumes. If humans were proportionately gifted, our conversations would be about as loud as a jet aircraft. • Spring Peepers are by no means mundane in appearance, but their small size and secret hideouts keep them from being seen by those

SIMILAR SPECIES

Western Chorus Frog, p. 202

Boreal Chorus Frog, p. 200

Gray Treefrog, p. 194

who refuse to crawl through their haunts. When discovered, Spring Peepers are perfectly delightful but surprisingly small—even after your second, third or fortieth discovery.

ID: dark "X," or less commonly a "Y," on back; very small; body colour varies from tan to grey; small toe pads.

LENGTH: up to 3 cm (1").

DISTRIBUTION: eastern Manitoba to Nova Scotia and Prince Edward Island. *Selected Sites:* Algonquin PP (ON), La Mauricie NP (QC), Fundy NP (NB), Kejimkujik NP (NS), Prince Edward Island NP (PE).

HABITAT: just about everywhere except in highly urbanized locations; brushy area; second growth forests close to permanent or temporary wetlands.

ACTIVITY PATTERNS: most active from late afternoon through the night; active from early spring to late fall, roughly the same time frame as the snow-free time of year.

REPRODUCTION: begin calling early in spring; while in amplexus the female lays 800–1000 eggs singly or in small groups attached to aquatic vegetation; tadpoles hatch in 2–3 weeks; transform into adults after 2–3 months.

FOOD: *Adult:* insects and other invertebrates. *Tadpole:* plants and algae.

CALL: breeding call is a single, loud, high pitched *peep* repeated up to 4000 times per hour! In large groups males may trill, which is intended to warn off rivals.

SIMILAR SPECIES: *Western Chorus Frog* (p. 202) and *Boreal Chorus Frog* (p. 200): 3 stripes down back. *Gray Treefrog* (p. 194): large, dark patches on body; "warty" skin. *Northern Cricket Frog* (p. 190): may occur on Pelee Island; dark triangle between eyes.

FRENCH NAME: Rainette crucifère

SIMILAR SPECIES

Northern Cricket Frog, p. 190

DID YOU KNOW? The Spring Peeper's species name *crucifer* means "cross" and refers to the "X" on the back of most individuals.

Boreal Chorus Frog

Pseudacris maculata

Canada's sloughs, ditches, and ponds are oases brimming with life and buzzing with activity, but it is the smallest of their amphibians that make them truly sing. From the parting of the ice in April through to the mosquito infestations of July, the Boreal Chorus Frog persistently announces its presence with its call. This simple and pleasing sound runs up the musical scale in evenly spaced sputters and signals to us that spring has arrived. • The sight of a Boreal Chorus Frog is a much more uncommon way to encounter an individual. Its simple physical form pales somewhat to its oratory, but nevertheless each individual is worth a look. The Boreal Chorus Frog provides a searcher with a visual challenge, as it refrains from calling when approached and sinks below the surface to merge in colour with the submerged plants and floating duckweed. With patience

SIMILAR SPECIES

Western Chorus Frog, p. 202

Spring Peeper, p. 198

Wood Frog, p. 188

and practice, however, it can be found if you know how to search. Many predators have mastered this technique, making the Boreal Chorus Frog one of the favourite meals of the marshes. The great variation in colour and design in this species ranges from browns to lime greens and spotted to striped, making every Boreal Chorus Frog one to remember.

ID: small; base colour ranges from lime green to dark brown; almost always a dark line running through eye and nose down along flank; 3 dark stripes (occasionally broken) down back; toe pads are not prominent.

LENGTH: 3–4 cm (1–1½").

DISTRIBUTION: from the Peace River area of British Columbia to southern James Bay in Quebec and north to Great Bear Lake in the Northwest Territories. *Selected Sites:* Elk Island NP (AB), Wood Buffalo NP (AB and NT), Prince Albert NP (SK), Meadow Lake PP (SK), Riding Mountain NP (MB).

HABITAT: still, murky water; woodland ponds, prairie sloughs, roadside ditches, flooded fields, beaver ponds, marshes, swamps, lakes and tundra ponds.

ACTIVITY PATTERNS: active from early spring to mid-fall; in harsher environments such as the arid grasslands, may enter a period of inactivity by late summer; tends to be more active from late afternoon to late night.

REPRODUCTION: begins calling day and night in very early spring; while in amplexus, female lays golf ball-sized egg masses on vegetation; eggs hatch within 2 weeks; tadpoles begin transforming into adults by early July.

FOOD: *Adult:* small aquatic and terrestrial invertebrates. *Tadpole:* plants and algae.

CALL: breeding call is similar to the sound of drawing your finger up the teeth of a comb; can be heard into late summer.

SIMILAR SPECIES: *Western Chorus Frog* (p. 202): best distinguished by its voice; ranges do not overlap. *Spring Peeper* (p. 198): "X" or "Y" on back. *Wood Frog* (p. 188): much larger; dorsolateral folds; less slender body shaping.

FRENCH NAME: Rainette faux grillon boréale

DID YOU KNOW? From the Canadian Shield, boreal forest, parkland, grasslands, Hudson Bay lowlands and mountains, the Boreal Chorus Frog occupies a more diverse range of habitats than any other Canadian frog. The Boreal Chorus Frog is tolerant of freezing, allowing its range to expand into the North.

Western Chorus Frog

Pseudacris triseriata

As an amphibian occurring solidly in central Canada and the Great Lakes states, the Western Chorus Frog's name just doesn't seem to fit. In spite of misgivings surrounding its name, the Western Chorus Frog is a genuinely pleasing animal to be around. It calls freely and loudly for much of the time the ground is thawed, but don't expect to find one so easily. Like other chorus frogs, this one is small and likes to spend its time where only frogs, mosquitoes and adventurous children like to venture. • The Western Chorus Frog is essentially the eastern counterpart of the widespread Boreal Chorus Frog. Fairly recently, evidence was brought

SIMILAR SPECIES

Boreal Chorus Frog, p. 200

Spring Peeper, p. 198

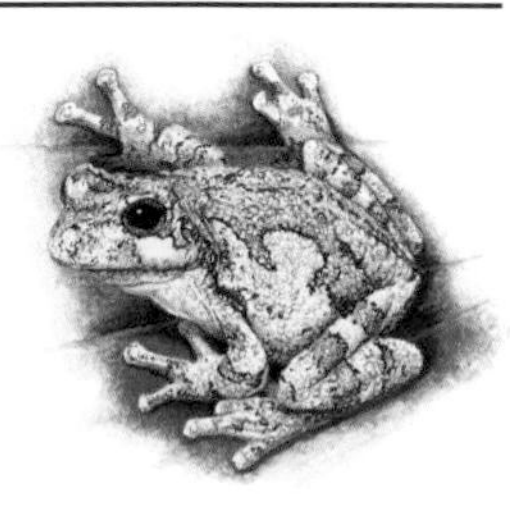

Gray Treefrog, p. 194

forward to the herpetological powers-that-be to split these two frogs into separate species. The Western Chorus Frog was the original one described, so it had priority on the scientific name *triseriata.* Don't be fooled by out-of-date books into thinking that the Western Chorus Frog was once much more widespread because of the vagaries of taxonomic changes. The Western Chorus Frog is still widespread in Canada, but its distribution is becoming more fragmented because of habitat loss.

ID: small; base colour ranges from lime green to dark brown; dark line runs from eye through nose, then down along flank; 3 dark stripes (occasionally broken) down back; toe pads are not prominent.

LENGTH: 3–4 cm (1–1½").

DISTRIBUTION: southern Ontario and along the Ottawa and upper St. Lawrence River valleys in Quebec. *Selected Sites:* Rondeau PP, Point Pelee NP, Luther Marsh (all in ON).

HABITAT: widespread; ditches, marshes, beaver ponds and lakes.

ACTIVITY PATTERNS: active from early spring to mid-fall; tends to be more active from late afternoon into late night.

REPRODUCTION: begins calling in very early spring; while in amplexus, female lays golf ball-sized egg masses on vegetation; eggs hatch in approximately 3 weeks; tadpoles begin transforming by mid-summer; frog lives up to 3 years.

FOOD: *Adult:* small aquatic and terrestrial invertebrates. *Tadpole:* plants and algae.

CALL: similar to running your fingernail up the teeth of a comb.

SIMILAR SPECIES: *Boreal Chorus Frog* (p. 200): ranges do not overlap in Canada; somewhat slower call. *Spring Peeper* (p. 198): "X" or "Y" on back. *Gray Treefrog* (p. 194): "warty" skin; irregular marks on back.

FRENCH NAME: Rainette faux-grillon de l'ouest

DID YOU KNOW? Western Chorus Frog tadpoles from the Toronto region were introduced to a pond in Newfoundland in 1963. The attempt to establish a population failed. That's probably a good thing, as most introductions mess up local ecosystems.

Glossary & Abbreviations

amplexus: the mating position of frogs and toads, with the male "hugging" the female

carapace: a turtle's upper shell

cloaca: a cavity at the end of the intestine into which both excretory and reproductive functions open

cranial crest: a bony ridge on the skull

dorsal: on/down the back

dorsolateral: between the side and the middle of the back

eft: the intermediate stage of a newt

extirpated species: one that no longer exists in the wild in a particular region but occurs elsewhere

herpetologist: one who studies reptiles and amphibians

herpetology: the study of reptiles and amphibians

introduced species: one that lives in, but is not native to, a region

invertebrate: an animal without a spinal column

keeled tail: flattened like a fin for better swimming

lateral: on/down the side

lungless salamander: any species belonging to the family Plethodontidae; in this book, p. 134–153

mole salamander: any species belonging to the family Ambystomatidae; in this book, p. 118–133

neoteny: the retention of juvenile features in the adult form

parotid gland: a salivary gland in front of the ear

plastron: a turtle's lower shell or breastplate

scute: each of the plates or scales forming a turtle's shell

spermatophore: a capsule containing sperm

tetraploid: an animal with four sets of chromosomes rather than two

tetrodotoxin: a poisonous substance found in certain species

triploid: an animal with three sets of chromosomes rather than two

tympanum: membrane covering the ear on a frog or toad

vagrant species: one that has wandered outside of its normal range

vertebrate: an animal with a spinal column

Abbreviations

CARCNET = Canadian Amphibian and Reptile Conservation Network

COSEWIC = Committee on the Status of Endangered Wildlife in Canada

COSSARO = Committee on the Status of Species at Risk in Ontario

KTTC = Kawartha Turtle Trauma Centre

NP = National Park
NWA = National Wildlife Area
NWR = National Wildlife Refuge
PP = Provincial Park

AB = Alberta
BC = British Columbia
MB = Manitoba
NB = New Brunswick
NL = Newfoundland
NS = Nova Scotia
NT = Northwest Territories
NU = Nunavut
ON = Ontario
PE = Prince Edward Island
QC = Quebec
SK = Saskatchewan
YT = Yukon Territory

Page numbers in **boldface** type refer to the primary species accounts.

About the Authors

Chris Fisher

Chris Fisher is a lover of all animals—furry, feathery, scaly and slimy. He received his B.Sc. in Zoology from the University of Alberta and also completed graduate studies there. Chris is a prominent nature writer; he has written many wildlife articles and field guides, including *Birds of Alberta* and *Mammals of Alberta*. He also gives lectures and presentations on wildlife and is passionately involved in wildlife conservation. Through his work, Chris aims to impart his appreciation for all things wild.

Amanda Joynt

As an ecologist, Amanda Joynt has enjoyed a varied career. Originally from the Okanagan Valley of British Columbia, she received her B.Sc. in Environmental and Conservation Science from the University of Alberta. During her university years, Amanda was a field technician for both Parks Canada and the Canadian Wildlife Service. After graduating, she wrote full time for Lone Pine Publishing. Later, she was able to travel throughout North America doing ecological surveys, including rare plant surveys in South Dakota. From 2004 to 2006, Amanda directed Children in the Wilderness Malawi, an environmental education and life skills program for orphaned children in the southern African country of Malawi. Amanda is now working as a biologist for the Department of Fisheries and Oceans in Inuvik, NT. She lives happily with her husband and dog.

Ronald J. Brooks

Dr. Brooks received his Ph.D. in Zoology from the University of Illinois. He has been a faculty member at the University of Guelph since 1970. Currently, he is Professor Emeritus. He is co-chair of the Amphibians and Reptiles subcommittee of the Committee on the Status of Endangered Wildlife in Canada (COSEWIC). He is also a member of several other committees for recovery of species at risk and is a member of the Ontario Endangered Species Assessment Panel. He has published over 125 peer-reviewed articles and hundreds of scientific reports on research for government and industry. Dr. Brooks is a recent recipient of the Mike Rankin "Distinguished Canadian Herpetologist" Award from the Canadian Association of Herpetologists.